International Conference on Smart Ship Technology 2020

Online
14-15 October 2020

ISBN: 978-1-7138-8306-7

Printed from e-media with permission by:

Curran Associates, Inc.
57 Morehouse Lane
Red Hook, NY 12571

Some format issues inherent in the e-media version may also appear in this print version.

Printed with permission by Curran Associates, Inc. (2024)

For permission requests, please contact The Royal Institution of Naval Architects
at the address below.

The Royal Institution of Naval Architects
8-9 Northumberland Street
London, WC2N 5DA
United Kingdom

Phone: 020 7235 4622
Fax: 020 7259 5912

publications@rina.org.uk

Additional copies of this publication are available from:

Curran Associates, Inc.
57 Morehouse Lane
Red Hook, NY 12571 USA
Phone: 845-758-0400
Fax: 845-758-2633
Email: curran@proceedings.com
Web: www.proceedings.com

International Conference

Smart Ship Technology 2020

14th -15th October 2020, Online Conference

PAPERS

THE ROYAL INSTITUTION OF NAVAL ARCHITECTS
8-9 Northumberland Street, London, WC2N 5DA, Telephone: +44 (0)20 7235 4622

The Royal Institution of
Naval Architects

INTERNATIONAL CONFERENCE

Smart Ship Technology 2020

14th -15th October 2020, Online Conference

Papers published in these conference Proceedings are peer reviewed.

THE ROYAL INSTITUTION OF NAVAL ARCHITECTS
8-9 Northumberland Street
London WC2N 5DA

Telephone: 020 7235 4622
Fax: 020 7259 5912

CONTENTS

UTILIZATION OF ARTIFICIAL NEURAL NETWORKS AS A BASIS FOR DECISION SUPPORT SYSTEM 95

S.Gökçay and A. Z. Saydam, Hidroteknik Nautical Design & Piri Reis University, Turkey
M.İnsel, Hidroteknik Nautical Design, Turkey
G. N. Küçüksu, Hidroteknik Nautical Design & Istanbul Technical University, Turkey

APPLICATION OF SMART TECHNOLOGIES TO SHIP DESIGN

A. Ramirez, R. Pérez, and J.A. Muñoz, SENER Ingeniería y Sistemas, S.A., Spain

SUMMARY

Smart technologies emergence in the field of the fourth industrial revolution makes it necessary to thoroughly analyse different aspects of ship design to find their best application, in order to provide value to the complete engineering process. Nevertheless, there is an underlying difficulty to find areas of application in this traditional industry, due to the highly stable production and design processes that have been achieved in Marine Engineering over the years. In this paper, different possibilities of applying disruptive technologies will be analysed, aiming to determine the advantages that these provide on diverse phases of the life cycle of the ship, and providing an explicit use case.

NOMENCLATURE

AI	Artificial Intelligence
AR	Augmented Reality
CAD	Computer Aided Design
CRISP	Cross Industry Standard Process for data mining
DA	Data Analytics
DL	Deep Learning
DM	Data Mining
KNN	K Nearest Neighbours
KPI	Key Performance Indicators
ML	Machine Learning
MVP	Minimum Viable Product
PCA	Principal Component Analysis
PLM	Product Lifecycle Management
PPL	Product Performance Lifecycle
RL	Reinforcement Learning
SL	Supervised learning
UL	Unsupervised Learning

1. INTRODUCTION

Smart technologies are nowadays very common topics as if everyone understood perfectly everything implied in the term. Truth be told, these are a compound of concepts, multiple technologies sharing the same final objective: digital transformation. However, the characteristics and differences among them require a profound and gentle analysis to determine how they can be applied to a specific industry.

The technologies that are being a key factor in the fourth industrial revolution are being consolidated over time. With regards to this particular situation, the marine engineering sector is forced into a slow and uncertain change process, stated as a must for its survival. The crucial point of this digital transformation is how to choose the best strategy, and how to measure your success on it.

A great difficulty at this point for companies is to maintain their strategies regardless of the general uncertainty due to COVID-19. Procedures may change, but strategy must be clear, consistent and stable to progress in the foreseen future.

When talking about digital transformation, several elements and exercises conform to the right strategy. To evaluate this process, it is necessary to evaluate the correct *Key Performance Indicators* (KPIs), which are particular to each process or business. Of course, profit-making must be one of these KPIs for success, but good practices should follow along the transformation process to achieve this goal.

According to McKinsey, less than 30% of the companies embarked on this innovative process succeed [1]. Focusing on expectations, there is a 45% chance to obtain less profit than expected [2]. This same study, see Figure 1, shows that applying different activities for digital transformation maximizes the possibility to overcome the expectations, in other words, improve the expected KPIs.

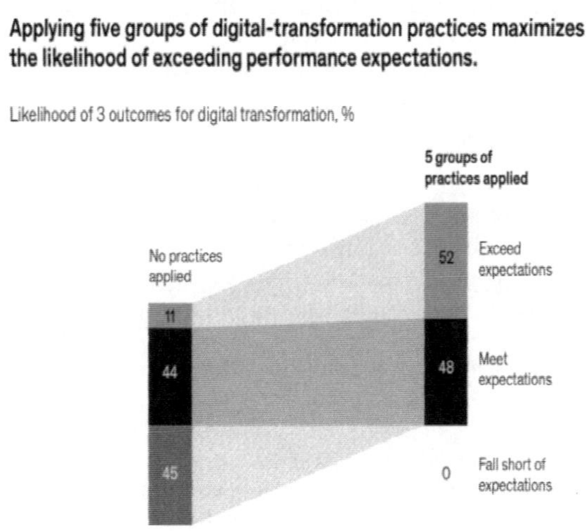

Figure 1: Results Improvement based on practices applied for the digital transformation [2]

Therefore, when talking about digital transformation on a company, not only is important to select the best technologies to introduce but how these are implanted. On the following Figure 2 extracted from McKinsey study, we can see the added likelihood of expectation-beating performance associated with each practice.

Figure 2: Impact of individual practices implanted during a digital transformation [2]

According to the previous figure, establishing clear priorities and agile processes are two of the key elements to truly make an impact on a digital transformation. Both these practices can be implemented by basing the transformation on *Minimum Viable Products* (MVP's), i.e. producing the minimum product or deliverables so that we can evaluate results on a short term and minimize failure. Selecting the right MVP will provide clarity of digital themes embracing agility.

The marine industry does not have an especially high margin of profit, but on the contrary, requires high investments and risks. Accordingly, investments must be carefully selected, assuring minimum risks. Having these in mind, a complete understanding of the technologies to deploy and the precise problem to solve must be achieved.

On the present paper, we will show different solutions to improve ship design, applying different technologies such as *Artificial Intelligence* (AI), *Machine Learning* (ML) and *Data Analytics* (DA). It should also be emphasised that these technologies are, according to Lloyd's Register, QinetiQ and Southampton University, highlighted for the short term in the Marine Industry [3].

2. STATE OF THE ART

In order to set the bases for the implementation of smart technologies in the ship design industry, specifically for the use case detailed in this paper; the following section explains the state of the art of some of these technologies.

The following image, Figure 3, shows the general understanding of different technologies and the relation among them:

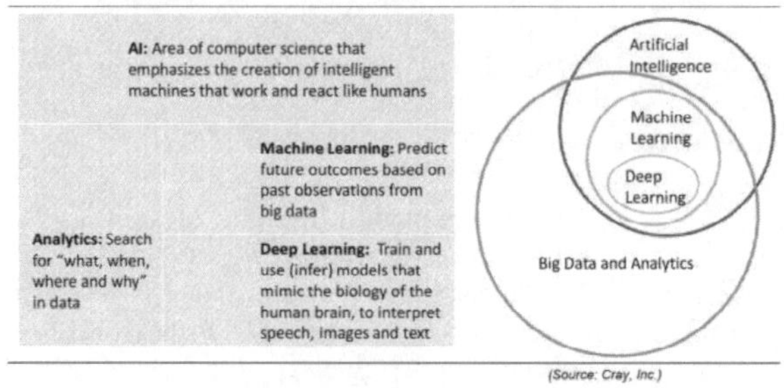

Figure 3: Relationships among AI, ML and DA [4]

2.1 Artificial Intelligence

Despite being born as the idea of creating machines capable of simulate human behaviour, AI has evolved to comprise multiple mathematic techniques used in computing and essential to create this "intelligent" machines, sharing a great number of concepts with data science.

2.2 Data Analytics

DA, also called *Data Mining*, is the discipline dedicated to analysing data to establish relations and provide conclusions.

The extraction and preparation of data are included in this field of study.

2.3 Machine Learning

ML allows the extraction of patterns from a data set. There are different types of analysis on this field, such as data clustering, support vector machines, association rule learning, Bayes algorithms, and all other algorithms included in *Deep Learning* (DL).

The Figure 4 explains in detail most of the analysis contained inside ML.

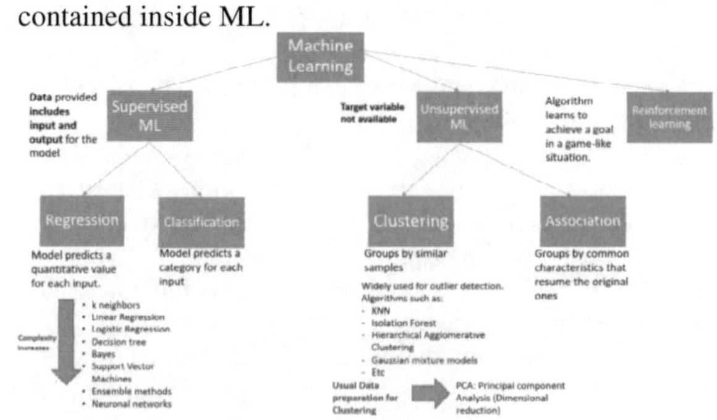

Figure 4: Machine Learning Components

In a simple way of explaining, ML is the evolution of ruled based systems, being able to distinguish among valid and surplus data, as well as deducing new rules not previously programmed.

Over the last five years, ML studies have focused on DL, high complexity algorithms that simulate human thinking. The Figure 5 shows the increase of studies on this field, due to the intrinsic interest of this field and the evolution of computing infrastructures and increase in data storage.

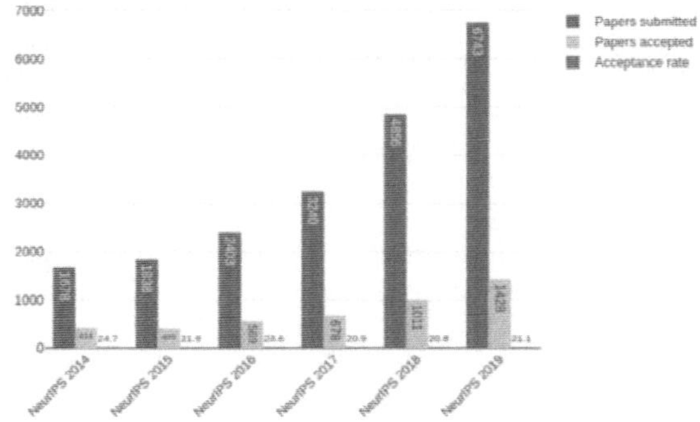

Figure 5: DL publications data [5]

3. BASIC CONCEPTS

The use case explained later on this paper aims to use ML techniques to determine the quality of CAD design. Inside this topic, we can find diverse techniques. Therefore, before explaining the specific use case, the following section details some related concepts.

ML concept is divided into three categories: *Supervised Learning* (SL), *Unsupervised Learning* (UL) and reinforcement learning (RL).

3.1 Supervised Learning

SL, see Figure 6, is a technique that, knowing the characteristic of the input data, provides an algorithm capable of establishing relations and identifying similar data. I.e. this system relates input and output parameters. Speaking in mathematical terms:

$$y = F(X)$$

Most problems considered in this category are classification or regression issues. Therefore, this type of ML is the most used at the moment.

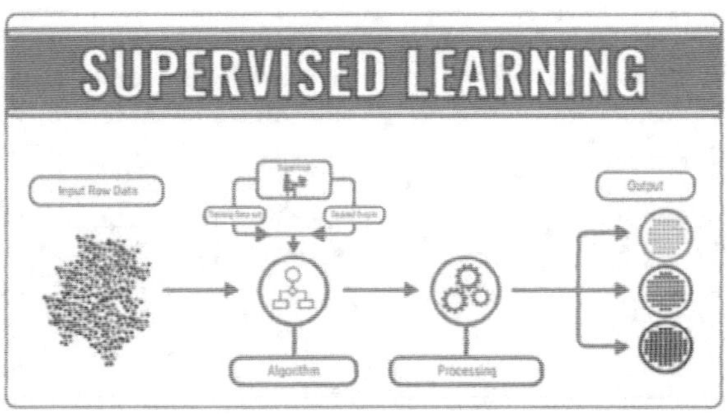

Figure 6: SL diagram [6]

The algorithm is capable of obtaining characteristic elements from the data to perform a classification according to the similarity of the extracted features.

The Figure 7 shows an example of SL used for classification. In this specific case, the input data are pictures of animals.

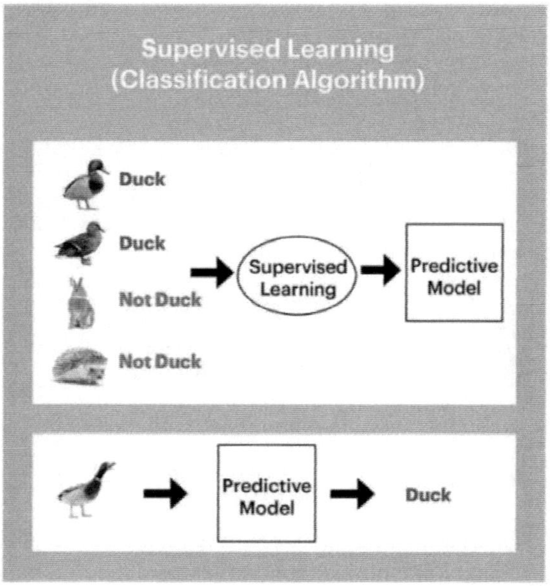

Figure 7: Example of SL [7]

The algorithm analyses the photographs and extracts characteristics among them, classifying and distinguishing in the end among those that contain ducks and those that do not. Of course, the more data provided, the better the training.

To train the algorithm, we must provide a relevant set of duck photographs, the model predicts and the supervisor corrects until results are adequate. In other words, the process would be like teaching a child to distinguish between photographs of dogs and cats.

In this specific area, the term "training" compounds the process to provide labelled data to the algorithm, making adjustments on it depending on the predictions it is providing. If, for example, in the previous figure the algorithm was trained with ducks and rabbits, it would be capable of telling us what animal appears on the pictures, not only if there is a duck or not.

3.2 Unsupervised Learning

UL, see Figure 8, does not have information about the categories or characteristics of the data but uses the algorithm to extract common features and group the input data in different groups or clusters.

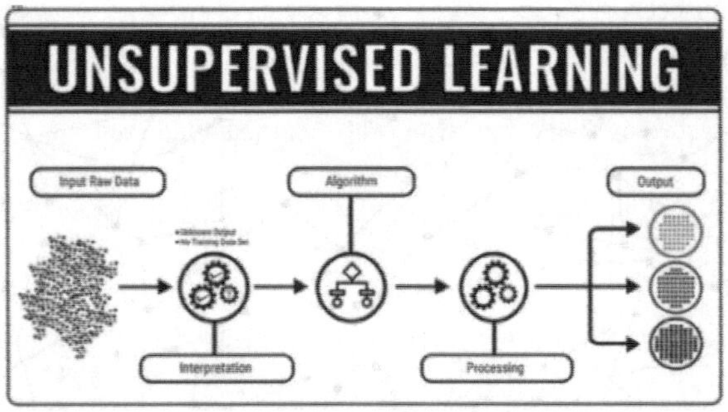

Figure 8: UL diagram [7]

The next image, Figure 9, shows a simple example of this type of algorithm. Without input information about the pictures, or training about types of animals, the system is able to group them based on similarities.

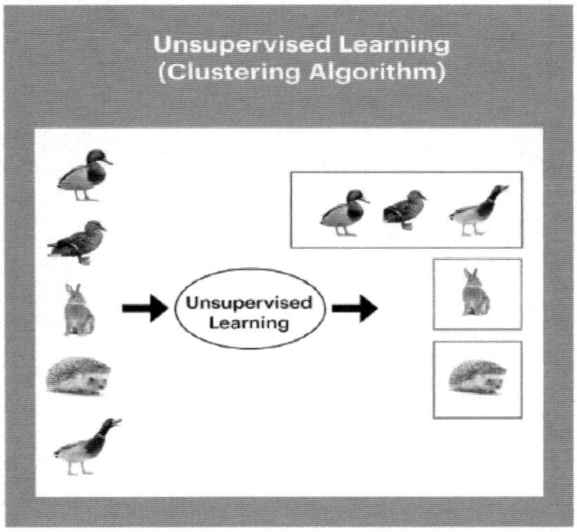

Figure 9: UL example [6]

3.3 Reinforcement Learning

RL is characterised by the fact that the algorithm is updated through a continuous trial-error and reward-reward analysis, seeking the optimization of the latter concept in order to find the optimal combination for certain external conditions, see Figure 10.

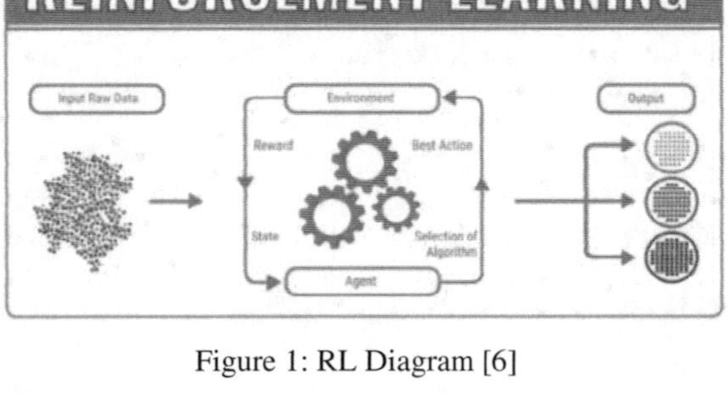

Figure 1: RL Diagram [6]

4. APPLICATION TO SHIP DESIGN

The application to ship design of the techniques included in the AI technologies requires the identification of use cases in which added value is provided to the design process. This additional value can be obtained through the application of technologies to different activities: automation of recurring tasks, standardization of objects or designs, identification of design anomalies, validation of designs, and so on. The main element that sustains all of the data.

As explained above, the use of these technologies requires the existence of algorithms, which are previously built with data. These can be generic, available for any industry or business; or specific, those corresponding to a particular industry, a shipyard or technical office. Due to the specificity of the marine industry, from our point of view, generic data will not be of much use for the design of particular shipyards or technical offices want to do. On the other hand, specific data of a design is not usually shared by shipyards and technical offices, since it is what gives them value and allows them to differentiate themselves from each other. This leads us to conclude that to have AI tools, the shipyard, office or naval designer, will need to create their own algorithms based on their own data.

Overall, it is not just about having any type of data. To apply this type of technology, it must be in abundance, accessible, and structured. The availability of sufficient data will also make it possible to identify processes or cases susceptible to improvement to which to apply the techniques explained above. In this sense, having a powerful design system with a relational and accessible database guarantees access to manageable data. In this specific case, as "data" we understand structural parts of a ship, equipment, supports, auxiliary structures, cables, connection nodes, cable and pipe routes, routes, weights, suppliers, prices, availability, construction strategies, logistics, etc. In short, all imaginable data that ship design requires from its initial conception to its construction. This input can be used for the design itself, controlling if the design fulfils the design rules, as well as production matters, assuring fewer errors, time savings and a wider standardization.

Therefore, to generate these algorithms we require previous knowledge of the concepts of ship design in order to find added value to processes; as well as specific data science expertise to select the best ML techniques for each problem. ML concept is often tried to apply in a general way, i.e. to any process. The reality, however, is that many processes acclaimed under this term are merely classical frequency studies or simple optimization problems, where there is no technology other than classic arithmetic. Due to this fact, to obtain algorithms that provide valid results, it is key to have a professional data analyst or technical profiles that can distinguish between different technologies and the best way to apply them.

One of the most used methodologies in DM is *CRoss Industry Standard Process* (CRISP). This technique divides the process into 6 main processes, with a partially flexible phase sequence. According to this methodology, see Figure 11, business and data understanding are key.

Figure 11: Diagram about CRISP Methodology [7]

In SENER we have used this methodology to develop the MVP explained below.

5. USE CASE: DETECTION OF DESIGN OUTLIERS

In the design of a ship, each technical office or shipyard uses certain practices, tools and strategies that aim to standardize the definition of the model. These try to reduce work time, minimize the use of materials, avoid errors and, ultimately, save costs. Any defined element that deviates from these work patterns are singularities or outliers that, although in some cases can be premeditated, in others they may be anomalies that cause an increase in design costs.

SENER has designed a group of ML techniques grouped under the term **FORAN Insights**. These techniques, applied to different stages in the life-cycle of the ship, detect singularities or outliers on the design, avoiding mistakes and terrible associated costs.

We consider anomaly or outlier an observation that differs deeply from the rest, and, hence, it may have been defined by a different mechanism [8].

The following sections are presented following CRISP methodology, as SENER studied them during the MVP.

5.1 Business Understanding

A key point in ship design, from an economic and production point of view, is to limit design errors as much as possible. Any small mistake in design phases can go unnoticed and reach production with shocking consequences for the project. Anticipating errors, preventing and correcting them in early stages has a cost up to ten times less than doing so in advanced stages of the production process.

Nevertheless, not all uniqueness is due to errors. It is also necessary to maximize the optimization on all parts, trying to standardize. Because of this, many shipyards and technical offices are implementing detailed design reviews in order to optimize production processes.

Taking these facts into account, the MVP developed by SENER uses ML algorithms on data set extracted from FORAN database, in order to detect outliers, singularities and anomalies on the design phase. As FORAN has a single database, supported by Oracle, the extraction of clean and organized data is simple. In particular, the MVP explained in this section has been focused on standard plates and profiles. The large number of elements used in this specific type of structure parts makes manual supervision extremely complex. With this type of approach, we provide a great impact on the project due to the volume of parts that can be affected.

5.2 Data Understanding

Internal structure standard parts (plates and profiles) are defines according to parametric criteria, defined by the designer in each case. One may wonder, what are the right criteria to consider a part as an outlier in this sample?

The most common answer to this type of issue is a compound of different type of analysis. Characteristics such as material or thickness can be analysed based on their frequency on the overall project. If, for example, we have just one plate on the project with a specific thickness, it is highly possible that this definition was due to human error. If it were a premeditated choice, it never hurts to be notified of this fact to be supervised or validated. On the other hand, evaluating the dimensions and shapes of the parts does not provide information because there is no defined pattern on which pieces are correct and which are not. However, analysis based on ML algorithms may be more useful in this regard. A piece of internal structure, for example, has parameters of different type and value. Something similar happens on profiles or reinforcements, which are also standard pieces. Therefore, we have data from which we want to extract common characteristics and outliers. The type of ML to apply is Unsupervised Learning. In our case, given that the data sample is very large, we can use clustering methods efficiently.

5.3 Data Preparation

UL algorithms will perform groups of data that are similar to each other in a two-dimensional space. Therefore, in the case of pieces with multiple parameters with different relevance to the set, we need to previously reduce their dimensionality. This is done with *Principal Component Analysis* (PCA), an attribute extraction technique that will create new dummy parameters for use in ML algorithms for more accurate results. Hence, the main objective of the PCA will be to reduce dimensionality without affecting the integrity of the data.

As input for the PCA it is possible to find a set of initial parameters ***p***, in order to obtain new non related ***q*** parameters, defined as principal components. These will have an associated weight that indicated their importance in the whole data set, as shown in Figure 12.

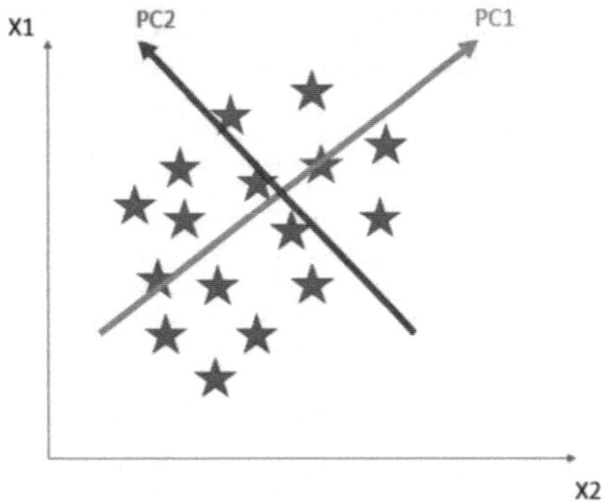

Figure12: Vector visualization of Principal Components

As an example, if we represent all principal components for a particular standard, we can visually see the similarities and differences among them.

On the following image, Figure 13, you can see represented in a 2D environment the principal components of all plates defined for a particular standard (A). The size of the points represented changes according to the number of equal parts.

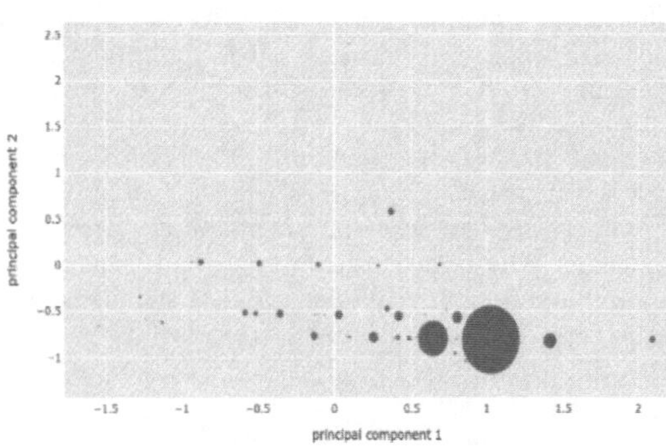

Figure 13: Principal Components Representation. Plates of A standard

The following picture, Figure 14, represents the same parts on a simpler way. We have marked in red colour those parts that, on a naked eye, seem to be design anomalies, as they differ tremendously from the data set.

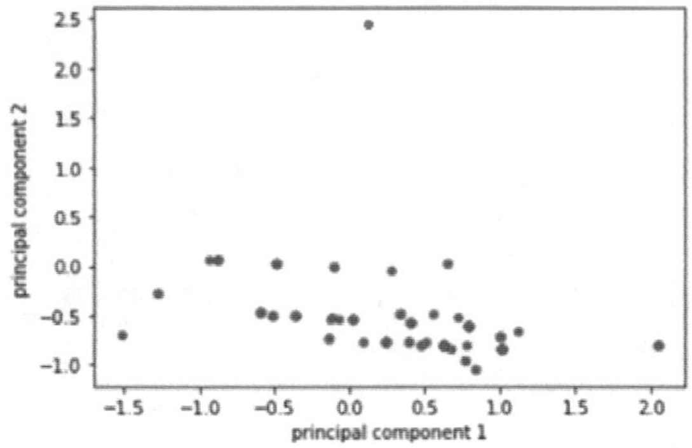

Figure 14: Detail Representation of Principal Components

One of the most important aspects in preparing the data to analyse anomalies is to detect those that are noise, see Figure 15. In other words, samples that are not general parts neither anomalies, but data that is not of interest for this specific study.

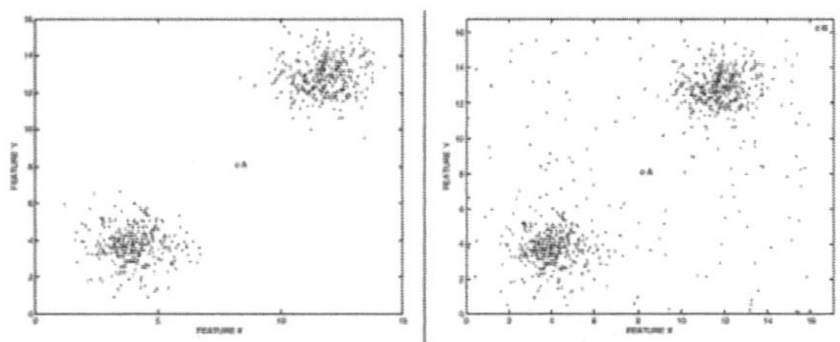

(a) No noise (b) With noise

Figure 15: Differences between anomalies with and without noise on data [9]

In the case of UL, where we do not have previous examples of anomalies for these data, noise is a complex factor. Generally, we will consider as noise those outliers modelled in a weak way, such as the limit data of the set. The different algorithms provide different metrics (scores) on the reliability of the result or the goodness of the value obtained. A true anomaly will generally have a higher anomaly score than data that is noise, although it is not a distinguishing factor. An expert's analysis is necessary to really distinguish between correct data and noise.

5.4 Modelling

In this task the aim to consolidate valid algorithms into a consistent model for the intended purpose. The model creation process is iterative, with phases of validation and updating of the different steps, as can be seen in the following diagram, Figure 16:

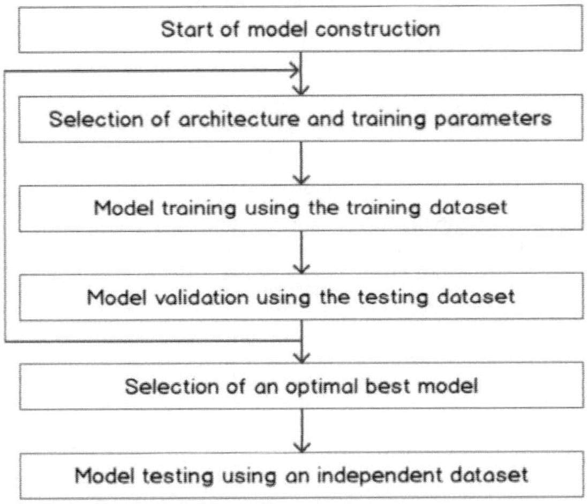

Figure 16: Diagram about evaluation of processes on ML [10]

After performing the PCA, different clustering algorithms can be used. To choose among them is key, being able to obtain erroneous results if the algorithm selection is incorrect. Each algorithm will make assumptions about what the “normal” behaviour of the data is, providing a score on its level of anomaly according to the quality of correlation between the data and the model. For this reason, an absolute understanding of the data is essential and representative values must be available to help with the task.

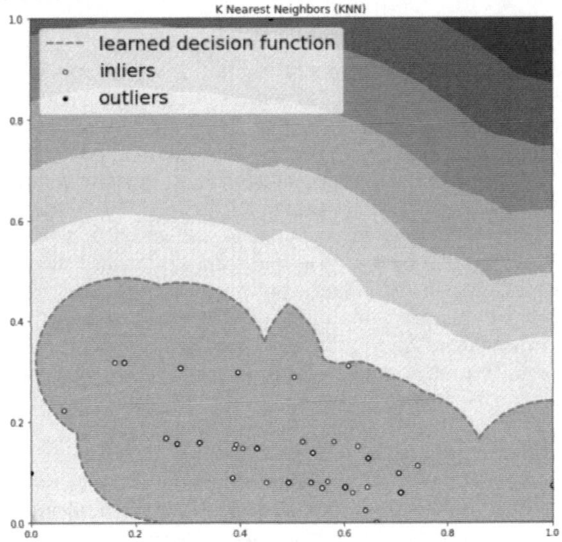

Figure 17: Example of detection of outliers in Figures 13 and 14 data

One of the significant values is the Z value, understood as the number of standard deviations by which a data point is distant from the mean. In this case, it is implicit that the data set is organized similarly to a normal distribution. In this particular case, we can state that a Z value equal or greater than three must be considered as an outlier [12]. In particular, for our example of standard plates, this type of value is not valid, since the data does not resemble a normal distribution in any case. Other common algorithms measure distances between points, where the outliers are at great distances from the bulk of the data. This is the case of the *K Nearest Neighbours* (KNN) algorithm, which measures the distance between each point and the closest centroid. This algorithm will obtain the number of clusters indicated by the user. There are different techniques to determine the optimal number.

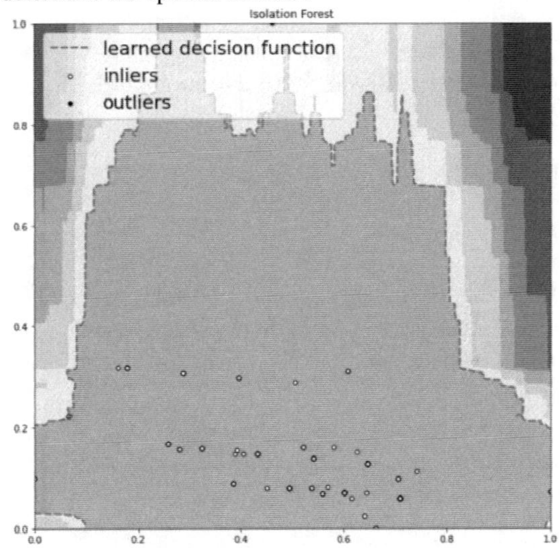

Figure 18: Example of Isolation Forest Analysis of Figures 13 and 14 data

Other common method for anomaly detection is Isolation Forest, an algorithm that builds decision trees based on data. The great advantage of this method is that it does not use distance measurements between the points. It is, however, of greater complexity.

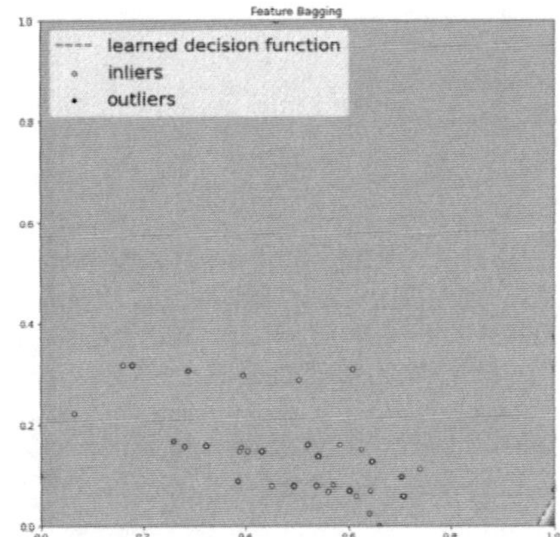

Figure 19: Feature bagging analysis of the data presented on Figures 13 and 14

Methods as Feature Bagging do not provide adequate results in this type of data sample. This type of algorithm aims to reduce the correlation between the different estimators within a data set, training with random samples instead of all the characteristics of the set. As can be seen in the Figure 19 on the specific example of the data presented in Figure 13 and Figure 14, according to this method we would not find anomalies in the data.

Another algorithm with inadequate results in the study of anomalies in standard plates is Histogram-Base Outlier Detection (HBOS), where a histogram is represented for each parameter to be analysed (see Figure 20). The frequency of samples in a set is used as an estimate of the density. On our particular MVP, results noticed did not contribute significantly to define anomalies.

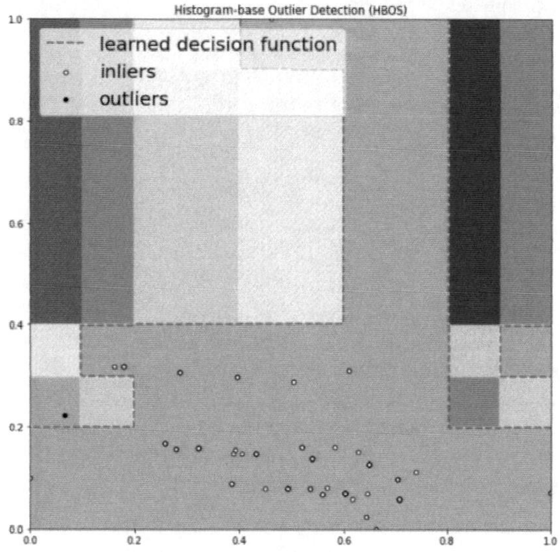

Figure 20: Results of HBOS algorithm on Figure 13 and 14 data

5.5 Evaluation

The next step is to evaluate the algorithms and select the appropriate ones for the data and the purpose of the analysis. In accordance with this and based on the studies carried out in SENER, as seen in the previous illustrations, we can affirm that the KNN and Isolation Forest algorithms provide very good results in the detection of anomalies in standard plates and profiles; compared to other methods such as ABOD (Angle Based Outlier Detection), CBLOF (Cluster-Based Local Outlier Factor), Feature Bagging, HBOS, Average KNN or Z Score. For other types of data, the models and results should be analysed to find the most appropriate ones.

In the example shown in the illustrations of this paper, thanks to this type of analysis, two anomalous pieces have been detected due to human error in the design, out of 386 of the sample for that specific standard. The process is autonomous and executable in seconds, nevertheless obtaining detailed results on the pieces to be analysed that, with a traditional method, would have gone unnoticed after hours of manual review.

5.6 Deployment

Finally, it is necessary to ease the analysis task based on the results obtained by the algorithms. Business Intelligence applications help to present results visually, through dashboards that facilitate understanding and decision-making in an agile way. These types of tools are, in a way, indispensable to present adequately and in a modern way the results of the ML algorithms that have been applied to the data. The following Figure 21 shows one of the "dashboards" obtained for data sets used in the MVP.

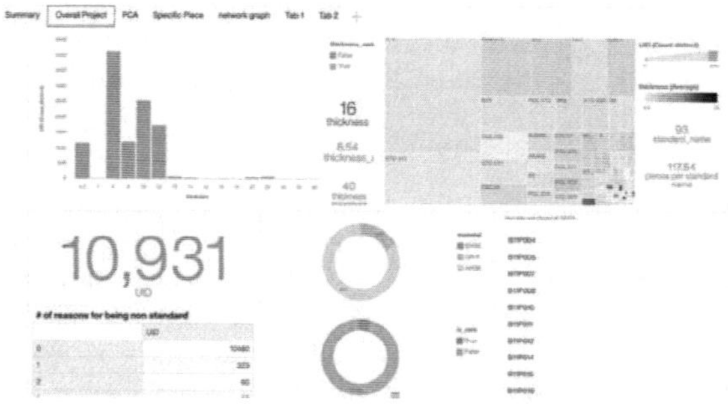

Figure 21: Dashboard for Outlier detection

6. CONCLUSSIONS AND FUTURE WORK

Digital transformation processes require appropriate strategies that allow to identifying the value gain obtained with the changes introduced and the investments made. In the context of current uncertainty and, in particular, in the shipbuilding business, a strategy that shows good results is to start with small projects that have sufficient viability to be evaluated without large investments (MVPs). It has been proved that this approach can provide results exceeding the expectations that were established at the beginning of transformation projects. Objectives must be clearly defined, stating the profit to be obtained and what are the technologies to use. Also, the teams of people who carry out this transformation must have adequate knowledge both of the business and the tools to use.

AI is one of the most disruptive technologies that can be used to achieve the digital transformation of business. In the field of ship design, it is possible to use ML to identify outlier parts to help the revision and standardization of the project. This article has made a descriptive analysis of the realization of an MVP, baptized as FORAN Insights. This project, using ML techniques on existing data sets already existing in a FORAN project, allows to identify possible anomalies in the design of standard parts of internal structure of a ship, quickly, efficiently and visually. This detection makes it possible to identify errors in the early stages of design. It also makes it possible to optimize the resources used, by ensuring the highest possible standardization of parts, standardizing materials and manufacturing, which results in the reduction of design and construction costs.

The MVP may be further developed, improving the parameters that allow users to identify and delimit the standard groups of parts. Besides, the system created must be further integrated with the FORAN project and with the design tools, allowing automatic and real-time detection to warn users of these possible anomalies.

The incorporation of visual tools for the presentation of results facilitates the work of engineers for decision making. These tools are complemented by dashboards that have to be adapted and personalized depending on the users or the shipyard needs.

7. ACKNOWLEDGEMENTS

The authors would like to express sincere gratitude to SENER for providing the support and information for this paper.

8. REFERENCES

[1]. MCKINSEY (2018). *Unlocking success in digital transformations*. McKinsey & Company.

[2]. MCKINSEY (2019). *Digital Transformation: Improving the odds of success*. McKinsey & Company.

[3]. LLOYD'S REGISTER et al. (2015). *Global Marine Technology Trends 2030*. Lloyd's Register, QinetiQ and University of Southampton.

[4]. MUÑOZ J.A. and PÉREZ R. (2019). *La integración de la Inteligencia Artificial en los procesos y metodologías de diseño naval*. Ingeniería Naval, no. 978, pp. 74-84, March.

[5]. MEDIUM. (2020). *NeurIPS 2019 Stats*. Available: https://medium.com/@dcharrezt/neurips-2019-stats-c91346d31c8f (Accessed 5th August 2020).

[6]. LOON, R.V. (2020). *Big Data Made Simple*. Available: https://bigdata-madesimple.com/machine-learning-explained-understanding-supervised-unsupervised-and-reinforcement-learning (Accessed 6th August 2020).

[7]. WIKIPEDIA. (2020). *Cross Industry Standard Process for Data Mining*. Available: https://es.wikipedia.org/wiki/Cross_Industry_Standard_Process_for_Data_Mining#/media/Archivo:CRISP-DM_Process_Diagram.png (Accessed 15th August 2020).

[8]. HAWKINS, D. (1980). *Identification of Outliers, Chapmal and Hall.*

[9]. AGGARWAL, C.C. (2017). *Outlier Analysis*. New York: Springer.

[10]. JOHNSTON, B.; JHA, A.R.; MATHUR I. and BATEMAN, B. (2020) *The Supervised Learning Workshop*. Packt Publishing.

9. AUTHORS BIOGRAPHY

Alicia Ramirez, MSc. in Naval and Ocean Engineering. holds the current position of Naval Architect and Marine Engineer at *SENER Ingeniería y Sistemas S.A.* She works at the Product Support Department of the FORAN System. Her previous experience includes product development and innovation in Naval and Industrial engineering. The Spanish Naval Architect Association (AINE) has awarded her for her final master thesis about an Offshore Patrol Cutter Vessel.

Rodrigo Pérez, Ph.D. is the head of Naval Shipbuilding Business Development in *SENER Ingeniería y Sistemas S.A.*. He is also professor in several subjects at the Marine Engineering School of the Technical University of Madrid.

Jesús A. Muñoz, Ph.D. holds the current position of FORAN System Product Manager at *SENER Ingeniería y Sistemas S.A.* He is responsible for the technical development guidance of the FORAN System. He is also professor of Applied Systems Engineering at the Marine Engineering School of the Universidad Politécnica de Madrid. He is author of several papers regarding to the trends and implementation of new technologies in marine design and shipbuilding.

IMPLEMENTATION OF STRUCTURAL DIGITAL TWINS IN MARINE APPLICATIONS

D. Chique, **H. Liu**, ABS, USA

SUMMARY

With the advent of digital technologies in ship connectivity, sensors, data analytics, and increased data availability, many efforts are being devoted to concepts and methodologies to enhance ship structural health assessment with vessel-specific data and analytics. This paper presents a structural digital twin framework that integrates vessel-specific data such as loading, navigation, environmental condition, and structural condition (corrosion and gauging) with first principles-based strength and fatigue assessment approaches. The framework denotes various tiers of the digital twin models applicable to different data availability levels, model sophistication and reliability, and potential benefits. Publicly available data sets, such as Automatic Identification System (AIS) data and weather hindcast data, rather than shipboard measurements, extend the digital twin framework to ships with limited onboard sensors. This framework is put into practice using a virtual hull monitoring approach through a containership case study. Combined with innovative inspection techniques and holistic data management, structural digital twins can provide potential benefits from data and data analytics. Proposed uses of the digital twin framework in the shipping industry are discussed, specifically for decision support for operations, drydock planning, inspection optimization, life extension, and fleet management.

1. INTRODUCTION

Vessel structural integrity management is a decision-making process that can be shaped by available data. Today, classification societies use data such as vessel age, type, and historical survey records to determine inspection requirements. A refinement of this process is the generation of a vessel-specific inspection plan. The plan is developed by including potential structural failure areas or critical areas identified through engineering analysis of the specific design. However, these approaches are still primarily based on statistical distributions of material properties, heuristic design philosophies, physical testing, and assumed similitude between testing and operational conditions [1]. In-service monitoring data, continuous structural condition assessment, and prediction capability can be integrated into the structural integrity management process to address these shortcomings.

A digital twin is a virtual representation of the physical asset, including its associated processes, systems, and information. As the actual physical asset changes throughout its life due to degradation, wear, and other processes, the digital twin is continuously updated using the physical asset data. By combining collected data with analysis models, a digital twin can be used to assess asset properties not directly measured, simulate the asset's response to various prescribed scenarios, or predict future conditions and failures based on forecasted operations and the asset's integrity management plan [2]. The digital twin for a vessel utilizes multiple data sources, including the vessel's design, structural models, , gauging and structural degradation, repair/modification/operational history, metocean data, inspection records, and in-service monitoring data. The data is maintained and analysed through the structural digital twin to assess and predict the structural condition, thus assisting structural integrity management decisions. Tygesen et al. present a comprehensive digital twin approach for fatigue re-assessment [3]. However, a comprehensive structural digital twin approach may not be suitable for all operators due to their various business goals and level of technical readiness.

This study presents a tiered structural digital twin framework with a balance of levels of fidelity and potential benefits with varying levels of data availability, analysis comprehensiveness, initial cost, and operating cost. The proposed digital twin framework integrates vessel-specific data such as loading conditions, navigational or operational data, environmental conditions, and structural conditions with first principles-based strength and fatigue assessment approaches. The tier level increases with increased data availability and in-depth analysis, resulting in a higher-fidelity structural digital twin with more potential benefits.

A tier-one solution as a case study using data from a container carrier trading along the Asia-North America route is used to put the framework into practice.

2. STRUCTURAL DIGITAL TWIN FRAMEWORK

The necessary components for a structural digital twin fall into three categories. The first is the representative condition models that reflect the current structural arrangement and conditions. These models may be continuously updated according to the physical asset's change using inspection and survey data to reflect the up-to-date structure's load-carrying capability. In most applications, the condition model is typically represented by a CAD or FE model. The second component is the experienced or forecasted structural loads, such as encountered wave conditions, cargo loads, loading patterns, and the vessel's operations such as speed and heading. The third is the data analytics using either physics-based or data-driven models to convert the experienced loads into structural responses, and then the potential failure and damage via the condition model. Finite element analysis and hydrodynamics are two commonly used data analytics approaches. Structural responses, such as displacement, deformation, and vibration, can then derive damage and failure modes in terms of yielding, buckling, and fatigue. The structural digital twin framework focuses on stratifying the different possible solutions by how well the models represent the vessel's current condition, how well the hull monitoring solution represents the experienced loads, or how reliably and accurately the data analytics converts the loads into structural responses.

Figure 1 summarizes the proposed structural digital twin framework. The first tier of the structural digital twin framework for vessels considers the experienced wave conditions and the operational history (draft and speed). The wave-induced global hull girder loads are derived through hydrodynamics or similar analysis and compared with those derived under design wave conditions as a risk indicator of wave-load induced potential structural damage. The second tier adds on a more reliable structural condition assessment, such as FE-based approaches, for a more accurate and holistic structural response estimation. The entire structural model will be solved, and the structural details will be evaluated against failure criteria periodically or as needed using the vessel's up-to-date condition models for enhanced accuracy. The third tier uses high-fidelity data, such as in-service monitoring and measurements from sensors to calibrate and validate the structural condition assessment achieved by Tier2. With the high-fidelity data, the Tier3 assessment can be more reliable. A Tier4 solution incorporates the digital twin with a risk-based framework, which fully considers the uncertainties relevant to the data and data analytics and the impact on the vessel's safety and operation caused by the potential structural damage or failure. A Tier4 solution quantitively estimates the risk and can provide decision support based on the probability of failure along with a confidence level.

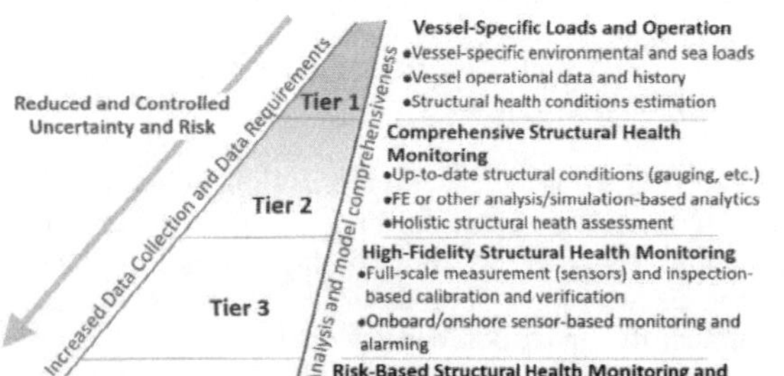

Figure 1: Structural Digital Twin Framework

In the current framework, the increasing tier levels reflect a reduction of the overall uncertainty and better controlled risk through a more accurate condition model, increased data volume and data quality, and more comprehensive and validated data analytics.

2.1 WORKFLOW

To form a complete structural digital twin, the three categories (condition model, structural loads, and data analytics) may be integrated and connected with the workflow shown in Figure 2.

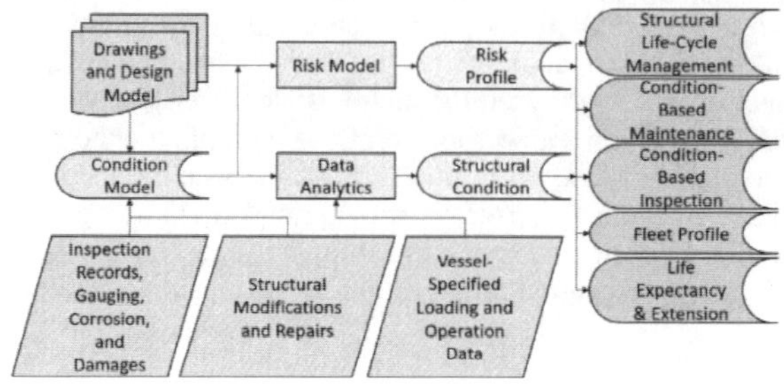

Figure 2: Structural Digital Twin Workflow

The condition model is typically constructed during the design stage and continuously updated with inspection and survey data, and structural repair and modification history. Data analysis uses vessel-specific loading, operational history, and the up-to-date condition model to derive structural responses and conditions. The risk profile is originally generated according to the specific design and design analysis and continuously updated using the condition model. The overall risk of the structural failure is constructed through the assessed and predicted structural conditions with the consideration of various accuracy and reliablity out of the implemented digital twin tier (the likelihood of failure and damage occurrence) and the risk profile (the potential consequence on the vessel's safety and operation). The structural digital twin can then be implemented to assist risk-based decision making such as condition-based inspection and maintenance, fleet management and operation optimization, and life extension.

2.2 UNCERTAINTIES

The uncertainties associated with a structural digital twin involve the collected data, condition models, and data analysis approaches. The systemic uncertainties include:

- The possible mismatch between the assumed design loading and design condition, and the vessel experienced loading and the degraded structural condition; and
- The intrinsic uncertainty of the analysis approaches.

Tier1 and Tier2 approaches seek to reduce the loading and model relevant uncertainties by introducing vessel-specific loads, operations, and up-to-date structural condition models to reduce the data mismatch. The vessel-specific loads and operations may come from the direct load measurements and the collected operational data. Alternatively, through publicly and commercially available data, such as vessel positions through Automatic Identification System (AIS) and weather hindcast, the vessel experienced wave conditions can also be derived.

Another source of inaccuracy comes from the approximation inherent in the use of digital condition models, and the data analytics and engineering analysis approaches. Using sensor data to calibrate and validate the digital models and analysis can help to reduce the model and analysis uncertainty and improve the assessment accuracy. A Tier3 approach seeks to reduce the data analytics associated uncertainties through high-fidelity data and validated models and analysis.

Random uncertainties, introduced by data noise, human error, improper treatment of the data and random operations pose challenges for assessing structural conditions with confidence for inspection or operation decisions. The random uncertainty and its consequence need to be well understood, estimated and controlled for safe operations. Risk metrics constructed using a risk-based approach with the relevant uncertainties considered is a feasible and practical way for risk and condition-based decision-making using a structural digital twin. A Tier4 approach addresses the details of random uncertainties, risk profile, risk assessment, and risk-based decision-making.

3. HULL MODELS

FE models are a standard tool in an engineer's toolbox for structural analysis. FE models are used in a structural digital twin to find structural hotspots or critical areas where small design margins exist and tend to have fatigue and strength damages. The model may be updated over time using current conditions by taking thickness measurements or tracking defects. Structural models are also typically accompanied by the supporting weight distribution models of lightweight components and deadweight components from loading conditions and compartments.

A study by Hulkkonen et al. explains the different components of a 3D FEM and how to use panel models to perform hydrodynamic calculations [4]. They show how to use hydrodynamic calculations to create stress transfer functions that drive fatigue analysis. The same approach is applied in the ABS *Guide for Spectral-Based Fatigue Analysis (SFA) for Vessels* [5].

The ABS *Guide for SafeHull-Dynamic Loading Approach (DLA) for Vessels* uses hydrodynamic models for structural analysis by generating dominant loading parameters and the corresponding design waves and load cases [6]. Each load case contains a dominant load component containing static and dynamic parts, such as still water bending moments and wave-induced bending moments. Seakeeping analysis generates response amplitude operators (RAOs) for each dynamic part of the load component. The RAOs describe the amplitude of the response per unit wave height under the varying conditions of a relative wave heading and wave frequency.

4. HULL MONITORING

The traditional method of hull monitoring is through direct measurements by sensors. Hull monitoring systems using sensors typically include strain gauges, accelerometers, Motion Reference Units (MRUs), pressure gauges, and temperature sensors. However, the installation and maintenance of these hull monitoring systems can be costly and logistically challenging.

There are several approaches in the literature for cost-effective hull monitoring solutions. Some methods, such as those described by Zhu and Collette [7] and Mondoro et al. [8] use hull monitoring data from a measurement campaign to calibrate the models used in the design stage. These calibrated models can give better estimates for fatigue damage accumulation. These models also apply to sister vessels that share similar structural configurations. However, while this method avoids the long-term costs of a hull monitoring system, it is still necessary to install instruments on at least one vessel.

Another approach is to use virtual hull monitoring; publicly available datasets such as Automatic Identification System (AIS) data allows tracking a vessel's location across the globe. Weather hindcast data such as those provided by the United States National Oceanic and Atmospheric Administration (NOAA) and the European Union's Copernicus program provides the wave environment information at the vessel locations. Finally, physical models from the vessel design stage allow the estimation of structural response to the varying wave environments. The approach presented in this study for virtual hull monitoring uses concepts described in the ABS Guide for SafeHull-Dynamic Loading Approach (DLA) for Vessels and the ABS Guide for Spectral-Based Fatigue Analysis (SFA) for Vessels.

Several industry studies follow a similar approach to virtual hull monitoring. A study by Thompson uses hindcast data to build wave spectra and a hydrodynamic model and FE model to generate stress transfer functions. They compared the calculated stress spectra from hindcast to those derived from full-scale measurements to show that virtual hull monitoring using wave hindcasts accurately recreates stress distributions [9]. Another study by Hulkkonen et al. again follows a similar approach for estimating fatigue life and concludes this approach may be valuable as a first-pass measure for estimating a ship's service needs [4].

5. TIER 1 EXAMPLE

A Tier 1 structural digital twin solution is implemented using the virtual hull monitoring approach and first principles-based strength and fatigue assessments for a large container carrier on the Asia-North America route. This virtual hull monitoring is then compared to a direct measurement approach using sensors to validate the accuracy.

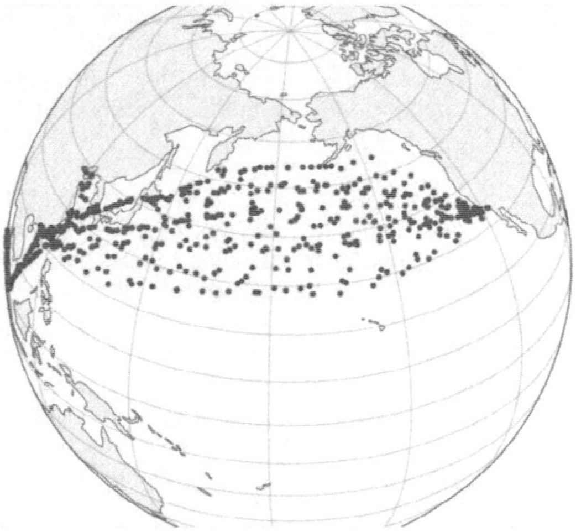

Figure 3: Vessel positions from 2006 to 2010.

5.1 OPERATIONAL MONITORING

Using AIS and Hindcast data, it is possible to monitor vessel operating parameters. This case study uses hindcast data from the publicly available hindcast database from NOAA WaveWatch III. A wave scatter diagram (WSD) is constructed for the vessel during the service period, which shows the distribution of the encountered wave conditions. A wave rosette is also derived using the vessel heading and wave direction information to show the vessel's probability of encountering a wave from a specific direction. An operational modes plot shows the distribution of speeds and drafts the vessel has experienced.

Together, these plots summarize the vessel-specific wave and operational conditions. This type of visualization allows for a quick understanding of the vessel's operating parameters, the difference from design condition and design assumptions, and a more straightforward comparison of different vessels' operations across a fleet.

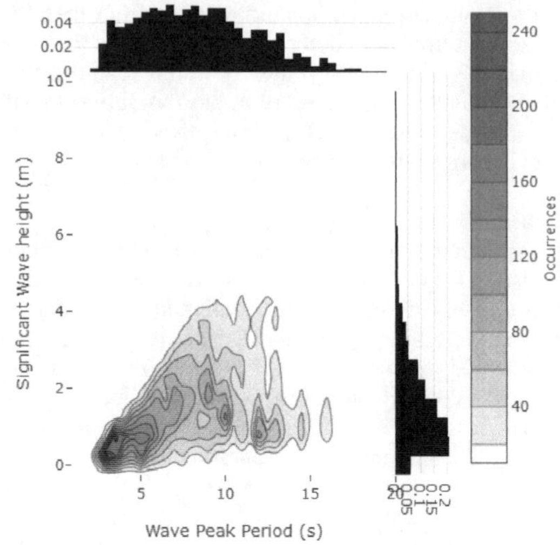

Figure 4: Vessel-specific wave scatter diagram.

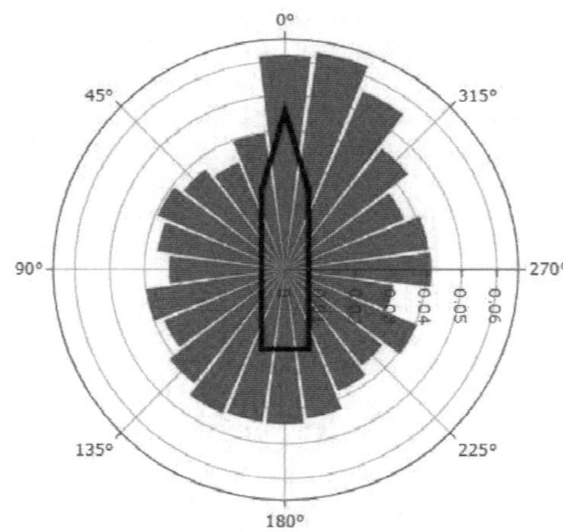

Figure 5: Vessel-specific wave rosette

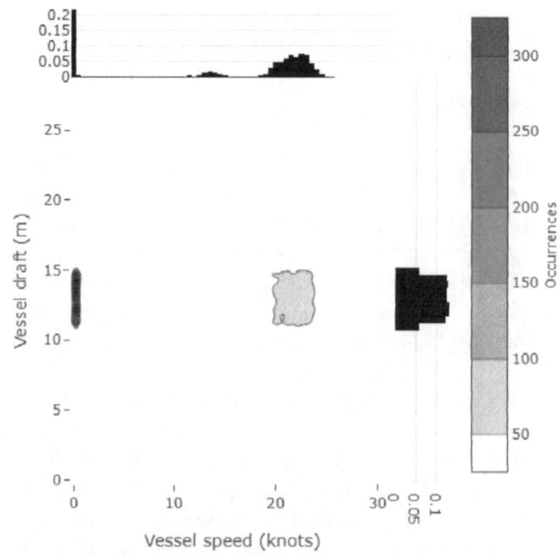

Figure 6: Operational profile

3.2 FATIGUE MONITORING

Two studies were identified that implement the virtual hull monitoring approach for fatigue analysis. For this specific vessel in the case study, Wang et al. implement a virtual hull monitoring for fatigue analysis [10]. The study compares the fatigue damage estimation using vessel-specific wave conditions to design wave conditions. The study found that using vessel-specific wave conditions resulted in significantly less fatigue damage accumulation than using design wave conditions.

Wang's methodology for performing actual fatigue damage estimation via the virtual hull monitoring approach follows the ABS SFA guide. The first step is to use the AIS and hindcast data to build a vessel-specific wave scatter diagram and wave rosette. The next step is to determine the stress transfer functions using FEM-based SFA analysis. The final step is to select which areas to monitor or the critical areas. Fatigue critical areas are design specific features that are areas of high-stress range or areas for which historical information for the vessel or the sister vessels indicates systemic hull-related findings.

As new AIS and hindcast data becomes available for the vessel, the fatigue damage estimates on these critical areas are periodically updated. Continuous fatigue damage monitoring can support the decision-making by owners and operators on operational management, fleet usage balance, inspection and maintenance optimization, as well as potential life extension.

5.3 STRENGTH MONITORING

While most studies following a virtual hull monitoring approach focused on fatigue damage estimation, strength monitoring is also an essential component of structural health. A proposed method for hull strength monitoring is to leverage concepts from ABS DLA.

DLA analysis finds strength-related critical areas by reviewing the FEM results for different load cases. Each load case contains both a static vessel loading configuration and a dynamic wave induced dominant load effect. Some examples of wave-induced load effects include vertical bending moments at midship, horizontal bending moments at midship, vertical shear force at various longitudinal locations, torsional bending moment at various longitudinal locations, and accelerations at the bow for potential slamming relevant damages. These dynamic wave-induced load effects are the dominant load parameters (DLPs) summarized in ABS DLA Guide for the common ocean-going vessels. With a linear assumption, each DLP has an accompanying RAO (Response Amplitude Operator) that describes the vessel's response to unit amplitude sinusoidal waves of various frequencies and headings.

Instead of monitoring the encountered wave, the proposed method for strength monitoring is to monitor the DLP loads obtained through the corresponding RAOs and the wave conditions encountered by the vessel for better correlation to the structural response and potential damage. It is important to note that the DLP RAOs depend on vessel operating conditions (loading and speed). Thus, if available, the ship's loading and navigation data can be used to select the appropriate set of RAOs. When this data is not available, assumptions may be made, such as using the draft and speed data from the AIS messages to represent the loading conditions, which will lead to less accurate estimation due to the ignorance of load distribution.

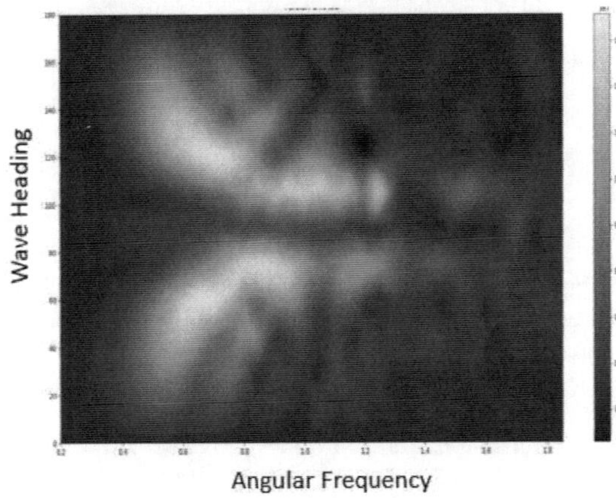

Figure 7: Torsional bending moment RAO.

Using the short-term extreme DLP response at each encountered wave system allows tracking the vessel's DLP responses. The ABS DLA Guide describes how to calculate short-term extreme DLP responses in an irregular wave system with an assumed Bretschneider spectrum. The short-term DLP response can then be put into context using the long-term DLP response using a referenced wave scatter diagram, such as IACS 34 that is typically used for design purposes with a 20 to 25-year design life. Expressing the short-term DLP response as a percentage of the long-term DLP response enables monitoring of the severity of encountered waves. This proportion serves as a risk indicator of the potential impact of the encountered wave on structural integrity. Various severity levels referring to the long-term reference extreme can be defined for possible triggered actions, such as self-inspection, continuous monitoring, and immediate actions, to avoid potential unnoticed structural damage and ensure the structural integrity and safe operations.

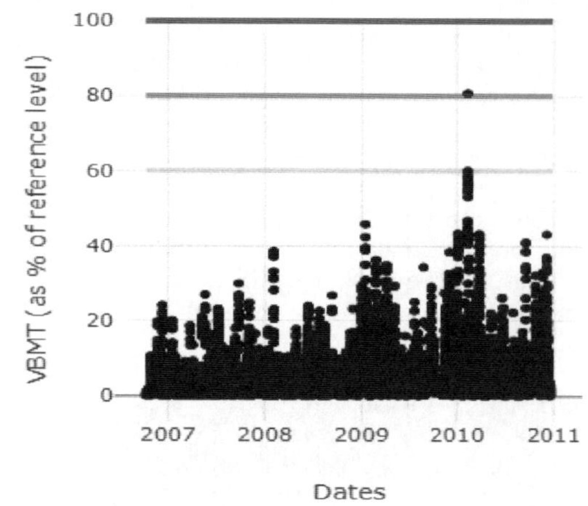

Figure 8: Vertical bending moment history from 2006 to 2011.

Figure 8 shows the monitored vertical bending moment amidship for the subject vessel in the case study. The vessel encountered an extreme wave event, which caused the short-term bending moment to go over 80% of the referenced long-term extreme. It should be noted that when the design wave and design condition are assumed as the reference long-term extremes, it is uncommon that the monitored DLP goes beyond the 100% reference. However, given all the uncertainties discussed in Section 2.2 as well as the uncertainties relevant to the workmanship and weld quality, a close to 100% reference monitored DLP may indicate a risk to structural integrity and proper preventative actions, such as inspections, may be needed to avoid unexpected structural failure.

5.4 COMPARISON TO SENSOR MEASUREMENTS

Several areas of uncertainty exist in the virtual hull monitoring approach, such as hindcast model uncertainty, hindcast model resolution, and ABS DLA approach limitations and assumptions. Despite the uncertainty, the approach provides valuable information to vessel owners and fleet managers on experienced wave-induced loads without additional sensors. Both Thompson and Hulkkonen conclude that this approach is valuable for fatigue damage estimation [4][9]. For this case study, direct sensor measurements are available and validate the virtual hull monitoring approach using DLP monitoring by showing good agreement.

Since the sensor data is time-series data, and the hindcast data is statistical data, the two datasets' comparison is not straightforward. A statistical approach makes for a better comparison by reducing the effects of measurement noise. For the sensor data, the spectral response's significant amplitude for the encountered irregular wave system is an appropriate figure for comparison. For the hindcast data, using the response spectrum, the response's significant amplitude *Rs* is given by:

$$R_S = 4\sqrt{m_0}$$

$$m_0 - First\ Spectral\ Moment$$

Equation 1: Significant amplitude of the response

This study uses the Vertical Bending Moment (VBMT) at midship for comparison purposes. The time-series VBMT measurements are segmented into three-hour intervals, assuming a wave event lasts three hours. This segmentation allows the calculation of the power spectrum for each three-hour interval. The final step is calculating the significant amplitude of the VBMT from the power spectrum. A single voyage from Asia to North America was selected to make the comparison and is shown in Figure 9:

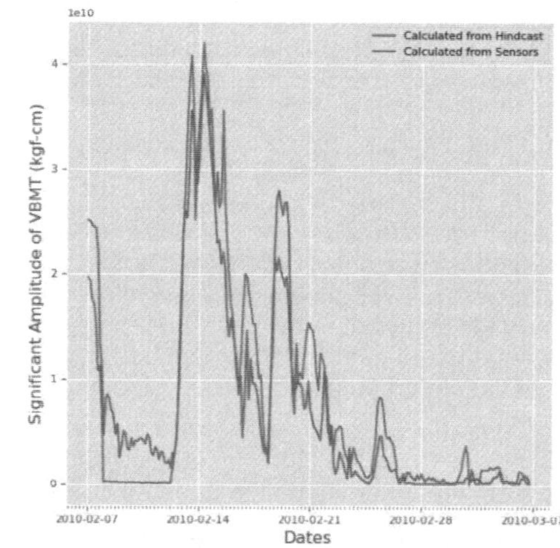

Figure 9: Sensor versus hindcast comparison.

The comparison results show that both datasets follow the same general trend and the virtual hull monitoring approach catches the wave-induced VBMT peaks well.

4. CONCLUSIONS

This paper introduced a structural digital twin framework offering different levels of fidelity and potential benefits. Shipowners and operators can adopt the proper tier suitable for their business goals, technical readiness, and financial status. The derived structural condition and failure and damage predication, combined with the risk profile and risk consideration, will support informed decision-making on inspection and maintenance planning, structural integrity management, fleet management, and potential life extension.

This paper presented a Tier1 solution using the virtual hull monitoring approach through a case study to put the framework into practice. The presented Tier1 solution provided a practical method of continuously monitoring global hull-girder loads and the risks of potential structural damage with publicly available AIS and hindcast data. It does not require any additional sensors and hardware onboard the vessel. The monitored global hull-girder loads were validated through comparison to direct sensor measurements. Using this AIS and hindcast data, an operational profile and encountered wave summary were constructed for post-voyage analysis and provide foundational data for future operational and design optimization. For a Tier1 solution, several fatigue studies already show how to perform monitoring through hindcast data. The reviewed studies' consensus is that the approach can provide valuable information for structural integrity management, fleet and operational management, and potential life extension.

Comprehensive structural hull monitoring through structural digital twin can provide owners, operators, and classification societies with important structural condition insights that can improve fleet management decisions and hull inspections and enhance safety, availability, and efficiency.

9. REFERENCES

1. GLAESSGEN, E. H., STARGEL, D. S., 'The Digital Twin Paradigm for Future NASA and U.S. Air Force Vehicles', *Structures, Structural Dynamics, and Materials Conference*, 2012.
2. SERRATELLA, C., LIU, H., and VANDERHORN, E., 'Supporting Structural Integrity, Coupling the digital twin with observed data', *Marine Technology*, October 2019.
3. TYGESEN, U.T., JEPSEN, M.S., VESTERMARK, J., DOLLERUP, N., PEDERSEN, A., 'The True Digital Twin Concept for Fatigue Re-Assessment of Marine Structures', *ASME 37th International Conference on Ocean, Offshore, and Arctic Engineering*, 2018.
4. HULKKONEN, T., MANDERBACKA, T., SUGIMOTO, K., 'Digital Twin for Monitoring Remaining Fatigue Life of Critical Hull Structures', *18th Conference on Computer and IT Applications in the Maritime Industries*, 2019.
5. AMERICAN BUREAU OF SHIPPING, *Guide for Spectral-Based Fatigue Analysis for Vessels*, 2016.
6. AMERICAN BUREAU OF SHIPPING, *Guide for SafeHull-Dynamic Loading Approach for Vessels*, 2018.
7. ZHU, J., COLLETTE, M., 'A Bayesian Approach for Shipboard Lifetime Wave Load Spectrum Updating', *Structure and Infrastructure Engineering*, 2017.
8. MONDORO, A., SOLIMAN, M., FRANGOPOL, D.M., 'Prediction of Structural Response of Naval Vessels Based on Available Structural Health Monitoring Data', *Ocean Engineering,* 2016.
9. THOMPSON, I., 'Virtual Hull Monitoring of a Naval Vessel Using Hindcast Data and Reconstructed 2-D Wave Spectra', 2020.
10. WANG, Z., VANDERHORN, E., LIU, H., SERRATELLA, C., 'Fatigue Damage Estimation and Prediction Using Vessel-Specific Wave Conditions', *SNAME Maritime Convention*, 2020.

9. AUTHORS BIOGRAPHY

Daniel Chique holds the current position of Senior Engineer at American Bureau of Shipping. He is responsible for helping class clients use data to enhance safety and efficiency. His previous experience includes hull monitoring systems, data analytics, and ship systems operation and maintenance.

Haitao Liu is a technical manager at American Bureau of Shipping (ABS). He is currently working on smart technology, structural digital twin, and systems engineering. His career in marine and offshore industry spans a range of disciplines, with highlights including advanced structural analysis, operational and environmental performance, and design optimization.

DEMYSTIFYING DIGITAL TWINS

R. Mansfield, J. Rigby, BMT, UK

SUMMARY

The 'digital twin' is now a recognised core component of the Industry 4.0 journey, helping organisations understand their complex processes, resources and data to provide insight into their business and help optimise their operations. Despite this there are still multiple definitions and understandings of what a digital twin actually is. All this has led to a "mysticism" around the concept. We at BMT define them as "A virtual representation of a product, asset, system or process". A collection of software models and simulations which is supplemented by data from live (or time delayed) measurements enabling an informed decision-making process. Twins are, first and foremost, learning systems, driven by data that's collected from sensors in real time. This paper seeks to demystify the different definitions and components of a digital twin; through this we will identify different examples of digital twins in practice or development today. The paper will also highlight the future growth of these systems, and the required development pathway.

1. INTRODUCTION

The digital twin is now a recognised core component of the Industry 4.0 journey, helping organisations understand their data, complex processes and resources to drive business insight and operational optimisation. But a digital twin still means different things to different people. Here at BMT we define Digital Twins as the following: "Digital twins are precise, virtual copies of machines or systems. These digital models receive regular data input from the physical system allowing for intelligent operational decision support." Twins are learning systems, driven by data collected from sensors in real time. Sets of complex digital models adapt to mirror all elements of a product, process or service. Powerful benefits garnered through efficiencies of an asset digital twins' product life-cycle management can be typically found in 4 areas:

Operations optimisation: Using variables like weather, fleet size, energy costs, or performance factors, models are triggered to run hundreds or thousands of "what-if" simulations to evaluate readiness or necessary adjustments to current system set-points. This enables system operations to be optimized or controlled during operation to mitigate risk, reduce cost, or gain any number of system efficiencies.

Predictive maintenance: In industry 4.0 applications, models can determine remaining useful life to inform operations on the most opportune time to service or replace equipment.

Anomaly detection: The model runs in parallel to the real assets, and immediately flags operational behaviour that deviates from expected (simulated) behaviour. For example, a petroleum company may stream sensor data from offshore oil rigs that operate continuously. The digital twin model will look for anomalies in the operational behaviour to help avoid catastrophic damage.

Fault isolation: Anomalies may trigger a battery of simulations to isolate the fault and identify the root cause so engineers or the system can take appropriate action.

Since the original inception of the twin concept in the eighties, great progress has been made across several sectors, typically with specific elements of a twin being implemented to achieve a specific aim for a specific system or sub-system. But, there is much to be done before the benefits of twins can be realised across industry on a wider scale. For instance, there are no standards in place for such tools, and there remain many challenges around open data capture, data storage, and performing effective analysis to provide decision support with confidence. There is not even a standard definition of a digital twin, with many variations in the accepted scope of a twin across sectors. To add further complexity to the picture, what started off as a niche engineering model is increasingly being expanded in scope and aligned to wider digital transformation initiatives. The growth momentum of the twin concept can be seen in today's digital world which encompasses a wide spectrum of technologies and market needs. There have been many interpretations and variations of the digital twin concept. It initially started with gaining momentum within high-value product manufacturing industries (mining, automotive and aerospace) but the concept is increasingly being found relevant across many other high-growth and fast moving consumer sectors including FMCG, Food and Beverage, Construction, Retail and Energy.

2. THE EVOLUTION OF DIGITAL TWINS

To provide enterprise wide benefit, a digital twin will need to be more than a single entity. Fully featured digital twins will be realised as a family of twins, each of which matures and grows in depth and value through its life. Models used to support asset performance will be different to those used in design and development, but there will be clear data linkages across all these model sets that need to be managed. Twins will both consume and create high volumes of data throughout their lives.
For simplicity we break down the lifecycle of a digital twin into three key stages and aspects (Figure 1).

i. **Design & Build**
ii. **Support**
iii. **Operate**

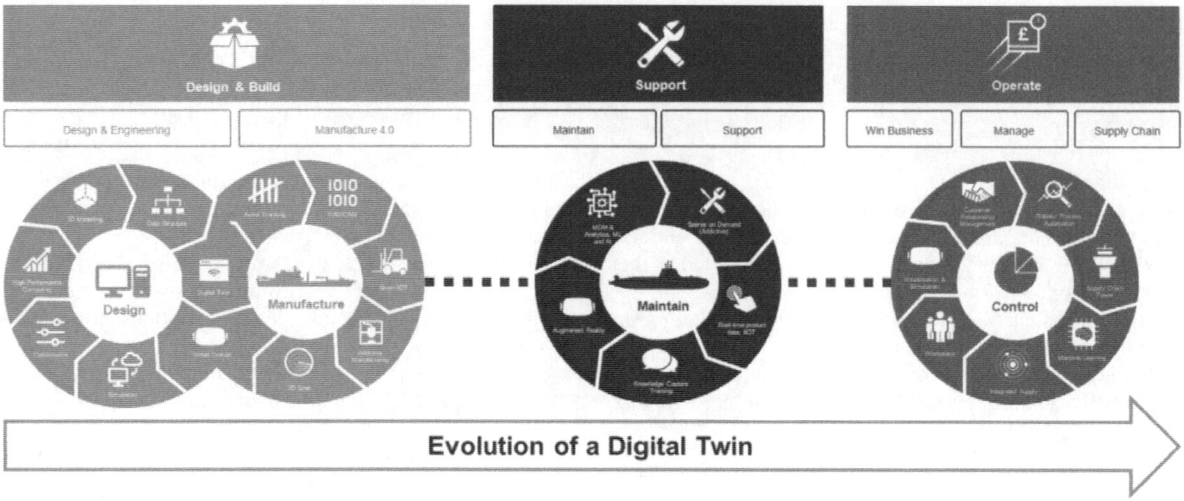

Figure 1- The three primary stages of digital twin evolution

In the design phase, a number of coherent twins will be created, from basic 3D geometric and design information through to complex high order simulation models and wider design information management. This 'Design & Build Twin' set will mature through build from being a data repository to becoming a manufacturing tool or "manufacturing or build twin" a key enabler for the future digital shipyard. This evolution will continue through to in-service where sensor inputs from the physical asset will enable the final 'Digital Twin' to be realised. This final digital twin will have two main faces, one from an industry support point of view to track and plan scheduled maintenance. The other from an operations point of view to control and manage the asset on a day to day basis and impact its operational decisions.

3. DIGITAL TWINS, WHY NOW?

Although the concept of a digital twin is not new (the term was first used in 2002 and the original information concepts date back to the 1980's), there are several enabling factors that are coming together to accelerate the implementation of twins. The first of these is cheap, high bandwidth sensing availability - IOT is a key enabler for twins. The second is the abundance of cheap data storage and advances in data management and mining. The third is the increased accessibility of the high-performance computing (HPC) required to manage the computing burden of a functioning twin. Finally, advances in integrated design and analysis environments are creating key digital threads at the start of the lifecycle, enabling the later creation of a twin.

4. CHALLENGES TO ADOPTION

Within the defence maritime enterprise many of the challenges to adoption come not just from the technical domain, but are equally a product of the commercial and supply chain environment. Defence maritime has traditionally contracted separately for design, build and maintenance contracts. These contracts are focussed on the provision of physical equipment and not on the provision of data and models. This approach typically results in the involvement of multiple prime contractors for each asset through life, and a correspondingly complex supply chain network. There is currently no mechanism to allow data and digital models to flow freely and securely through this enterprise, whilst protecting OEM's intellectual property. It is likely that a future digital twin for defence maritime will span multiple solution platforms, multiple companies, and multiple solution platform vendors.

5. BMT'S FOCUS

BMT sees digital twins as a key element of industry 4.0, having seen the benefits and savings they are bringing to our customers. BMT's Whole of Mine Digital Twin is just one of many bespoke examples. Built on our Remote Operations Automation and Robotics (ROAR) software platform and using lidar mapping capabilities, and expertise in 3D algorithmic decision making, our solution integrates precise Terrain data. Machine data, Health data and Production metrics into a single 3D immersive and intuitive interface. This allows production teams and planners to Measure, Monitor, Report and Optimise mining operations in real-time, reducing costs and risks.

BMT continues to develop and expand our digital twin technology into other submarkets, with ongoing projects delivering Autonomous TLO (Train Load Out) Digital Twins and Hopper Surface Mapping and Wagon Surface Mapping solutions for train unloading rail operations. Other technologies are in development for Stockyard Mapping and Ship Hold Surface Mapping for bulk materials handling customers in ports.

To support future maritime defence digital twins BMT is taking a multi-threaded approach through a range of internal strategic initiative and R&D activities. Key activities being undertaken include:

a: Digital threads in design, enabling the creation of the twin
BMT is mapping out our design toolset of the future for ship design, creating an integrated digital design twin, with advanced integrated simulation and analysis capabilities which enable the leftshift of risk from physical testing back into the design phase.

b: Integrity, security
End to end (sensor to decision) secure data transmission, management and verification; digital twin cyber threat detection, protection, response and recovery; a digital twin ship or fleet is an asset that is susceptible to threats that are analogous to those seen by physical ships.

c: Interoperability, supply chain engagement
Federated twin high level architectures and APIs that enable prime contractors and OEMS to support a digital ship with 'digital equipment' whilst protecting I.P. and ensuring interoperability of 'digital assets'.

d: Digital twin return on investment
Research into future intelligent asset management; a digital twin creates value when embedded into a holistic intelligent asset management system, supporting decision making and business model optimisation at portfolio and organisational levels. These solutions not only highlight BMT's expertise and abilities to develop bespoke customer solutions, they also reveal our commitment to understanding, embracing and leading digital transformation programmes within the industries we serve.

6. CONCLUSIONS

In conclusion, although there are many terms and phases being used to describe and talk about Digital Twins, we need to have clarity across industry about what we mean by a true "Digital Twin". The term has become more of a buzzword and if often used in places where it probably shouldn't. To this end BMT advises all its customers to utilise the following definition: "Digital twins are precise, virtual copies of machines or systems. These digital models receive regular data input from the physical system allowing for intelligent operational decision support." Without the intelligent operational decision support aspect it is not a "true digital twin".

Although they are not something to be afraid of you should be careful to understand the application and uses of your digital twin before information is collated. These information systems are incredibly complex and need to be carefully considered and constructed. They will of course evolve and grow over time, but the core information structure will need to remain the same.

Overall, we see Digital Twins as a key part of Maritime 4.0 and are working hard with our customers to turn the theoretical benefits it can generate into reality.

7. ACKNOWLEDGEMENTS

The authors would like to acknowledge all those who supported the digital twin research conducted at BMT and the production of this paper.

8. REFERENCES

1. S Willmore, J Rigby. "The Challenges and Opportunities for Developing Digital Twins to Support Maritime Operations", Marine Industry 4.0, 5th November 2019, Rotterdam, The Netherlands.

2. BMT, 'Digital Transformation – Building Security by design into your digital transformation project', BMT Focus Magazine July 2020.

3. G Michael, 'Origins of the Digital Twin Concept'. https://www.researchgate.net/publication/307509727_Origins_of_the_Digital_Twin_Concept 1 July 2020.

4. SSI, 'Managing Vessel Repairs with a "Digital Twin", https://www.maritime-executive.com/blog/managing-vessel-repairs-with-a-digital-twin, 4 Feb 2019.

5. Laurence Goasduff, 'Confront Key Challenges to Boost Digital Twin Success', https://www.gartner.com/smarterwithgartner/confront-key-challenges-to-boost-digital-twin-success/, 13 Mar 2018.

6. A. Parrott, Dr. L. Warshaw, 'Industry 4.0 and the digital twin', https://www2.deloitte.com/content/dam/Deloitte/cn/Documents/cip/deloitte-cn-cip-industry-4-0-digital-twin-technology-en-171215.pdf, 2017.

7. Jonathon Scott, "How does one get started with PLM and the Digital Twin", Machine Design, https://www.machinedesign.com/cad/how-does-one-get-started-plm-and-digital-twin, 1 July 2020.

A CONTEMPORARY APPROACH FOR TRAJECTORY PREDICTION OF AUTONOMOUS SURFACE VESSELS (ASV'S) BASED ON LONG SHORT-TERM MEMORY (LSTM) NEURAL NETWORK

D. Rajani, Dwarkadas Jivanlal Sanghvi College Of Engineering (DJSCE), India

SUMMARY

Artificial Intelligence techniques and algorithms have a cornerstone role in the rapid and efficient development of the maritime industry. The algorithms under this domain are extensively used to automate controls of sea-going vessels hence rendering their operations autonomous or in some cases remote-based from an onshore facility. These Autonomous Surface Vessels (ASV's) generate multifarious data from a number of different sensors on-board the vessel for self-estimation with respect to its environmental conditions and position. The fundamental aspect of any ASV is trajectory prediction and planning. This aspect has been explicitly delineated in this paper by using the data from the Automatic Identification System (AIS), which is a continuous automatic tracking system that provides us with information like the estimated vessel position and velocity, which is used for data pre-processing to fortify data probity. A Long Short-Term Memory (LSTM) recurrent neural network model, combined with the sequence prediction methodology, rooted upon the data obtained from AIS, is proposed to predict the trajectories of the vessels. The empirical results evince that the model illustrated can predict ship trajectories based on route data accurately and hence proves that the model is applicable for the efficient operation of autonomous navigation.

Keywords— Trajectory Prediction; Autonomous Surface Vessels; Automatic Identification System; Long Short-Term Memory (LSTM) recurrent neural network; Autonomous Navigation

1. INTRODUCTION

With the advent of technology computers have evolved in such a manner that these devices today are not only used to sustain tortuously programmed software or perform repetitive tasks. An anachronistic computer requires supervision, but today's computers contend humans by performing some byzantine tasks. These smart machines widely known as Artificially Intelligent systems which are backed with some spectacular state-of-art machine learning algorithms which not only render the memory of these machines to compile intricate problems quickly but also allocate them with a certain level of autonomy.

The concept of Artificial Intelligence (AI) is substantially applicable to all the facets of industries thriving in this world in order to forge ahead novel ideas for efficient, robust and accoutre in all a generalized approach to random world problems affecting, in particularly, to the industry under consideration. The application and development of burly and smart algorithms having forefront perspicacity is the need of the hour for indeed the most protean industry, i.e. the transportation industry. The maritime industry is the one which seems luddite in incorporating such technology in its core technical system.

Seafaring vessels have a huge role in sustaining world economy. As stated by the International Chamber of Shipping (ICS) in their recent survey, maritime vessels account for 92% of all international trade making them, the "Marrow" of the world economy. But after all these boats are also some buoyant petri dishes prone to unpredictability. In order to extirpate the fundamental threats to autonomous vessels, like reliability, piracy and bad weather conditions, autonomous navigation can, to certain extent, resolve all these threats by augmenting AI models with traditional navigation.

The novel approach posited in this paper provides a nuanced foresight into the subsequent position of the autonomous ships keeping in mind the transient environmental conditions which affect the overall trajectory and speed of these ASVs. This approach can be used to predict ASV's position well before in time for collision avoidance or to provide efficient and timely rescue operation at sea (commercial purposes) without any kind of fallacies in the interpretation of the data.

From scientific view point ASVs are widely used in oceanographic research, from engineering view point these vessels are used for survey of offshore wind farms, above-water bridges and offshore oil rigs and also these vessels are highly being used by naval services for mine-sweeping purposes. From all these applications mentioned, ASVs generate a multitude of data which is directed either to the satellites or to the onshore remote controlled facilities. For monitoring ASVs in highly stochastic conditions, the beforehand details are pertinent which can be used to predict the subsequent course of the vessel.

The data generated must be computationally compatible to process and this requirement is fulfilled by using the data-set received from Automatic Identification System (AIS) .The AIS is an automatic tracking system that uses digital VHF radio-based transponders on-board vessels to transmit and receive data in form of signals. Satellites or onshore transceivers are used to detect AIS signatures, provided by AIS sensors, such as distinct identification, coordinates, heading, and speed, can be mapped on an electronic chart display. AIS is intended to assist maritime remote sensing facilities to track and monitor vessel movements. However, due to varied reasons there might be many kinds of signal losses due to range constrictions or most probably due to environmental constraints, these factors lead to improper data updates which may falsify the overall output for trajectory determination. Hence these new AI based techniques are used to compensate for the loss of data transfer intermittently.

According to reference [1] the most predominant method of path prediction is used by employing Kalman filters, which uses an algorithm that uses a sequential form of measurement which is reshaped by statistical noise. Based on this reshaped data an Autocovariance Least-Squares (ALS) approach is applied which produces an estimation of unknown quantities, hence predicting vessel trajectories.

In reference [2], the proposed model is the Gaussian mixture model which uses state based calculations which can precisely predict handful trajectories, but here the main drawback is that this method is model based and not data based i.e. for each trajectory we have to construct distinct mathematical versions, hence this method is highly inefficient for navigational purposes.
The method in reference [3] manipulates Markov chains of second order and feed-forward based network using data accumulated from Global Positioning System (GPS). The accruing probability helps in determining the future data points with great accuracy but this leads to increasing hidden layers accounting for data warehousing problems and spatial costs.

The technique in reference [4] utilizes Bayesian model to solve low trace and poor prediction drawback. Primarily, a Markov model is implemented to correlate consecutive coordinates and then the trajectory is divided into small samples to predict the subsequent position, but the drawback here is overfitting due to excessive clustering complexity.

In addition to the above methods there are many commonplace machine learning algorithms which can produce more or less an accurate prediction but these need to be further researched upon to create a canonical model for real-time data. Rooted on the vessel's present path in reference [5] Artificial Neural Network (ANN) is used to fit this current trajectory and to foresee upcoming trajectory. The statistical analysis proved that ANN has a robust path prediction and model fitting than other methods. This results form the basis for selection of LSTM, a more neoteric approach which can be further tuned to provide a more accurate and generalized result.
The LSTM model, a modification of Recurrent Neural Network (RNN) has itself shown greater decisiveness with definite results in limited time without any complex clustering. A common prediction algorithm uses RNN to predict data as in [6].The predicted value exclusively depends on the historical values taken as input. The only problem with RNN is the vanishing gradient problem which impedes learning rate of the model. The gradients transfer information in RNN to upgrade parameters, and when the gradient converges to minima, the parameter updates tend to be negligible, as if no learning has occurred.

Therefore, a coherent approach is posited in this paper wherein we train a LSTM model for sequential predictions from the AIS data-set. LSTM is an order dependent sequence prediction approach which is a modified form of RNN, which utilizes control gates allowing it to keep or forget the information over the course of data modeling causing back-propagation and hence solving the gradient descent problem without any additional clustering.

2. LONG SHORT-TERM MEMORY (LSTM) RECURRENT NEURAL NETWORK

LSTM is an artificial recurrent neural network which is quite pertinent in development of statistical models. It is basically is a recursive category of neural network which is used for processing and computing long sequential information and also can be a great compensation in case of real-time data problems like information lag or signal attenuation. LSTM was put forward by Hochreiter & Schmidhuber in 1997.

The LSTM is a sub-set of RNN. A RNN has three fundamental layers viz. The input layer, hidden layer and output layer. The conventional neural networks have all the inputs and outputs are not the functions of each other, but in case of RNN where predicted values are to be determined in long sequential data, the input values with respect to the hidden layer inputs are required to be memorized till the final layer. The core of RNN is the hidden layer, which memorizes the previous values.

A common RNN cycle involves; firstly, the input values get reshaped into compatible vectors. Then the RNN processes this sequential data bit by bit. During processing, it conveys the previous hidden layer conditions to the next step of the sequence. The hidden layer acts as the network's memory which holds unseen details on past data.

The reshaped inputs and previous hidden conditions, in an individual RNN cell, are compiled into a vector, which has present and past information. This vector parses through the activation function, and the output is the new hidden state, or the memory of the network.

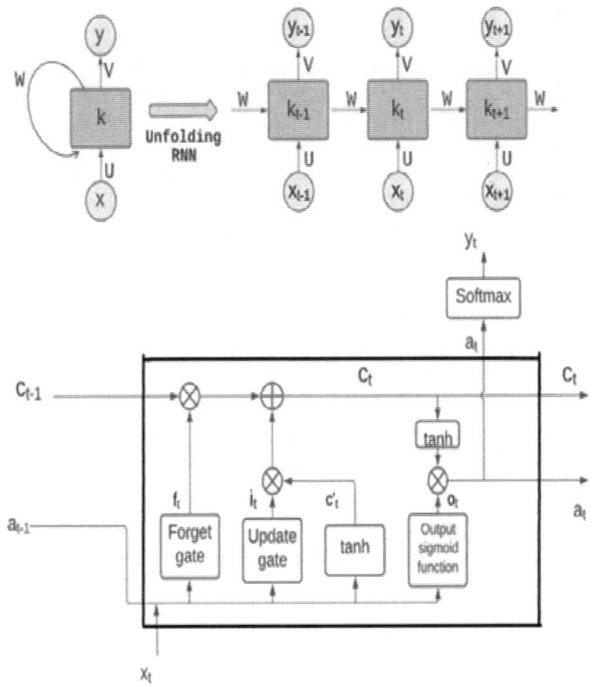

Figure 1: (1) Unfolded hidden layer of RNN; (2) LSTM memory block

Figure 1(1) shows the relative unfolding of the consequent RNN hidden layers. This figure visualizes the model as a multiple recursion transfer function in which each layer replicates each other activity based on the input received. As shown in the Figure 1(1),' t-1, t, t-2…'which are consecutively placed time series data-points. Here 'x' is the sample input and 'y' is the sample output and 'k_t' is the memory of the applied sample at time t. Equation $k_t = g(W*k_{t-1} + U*x_t)$. W signifies weight of the input, U signifies weight of the input sample at that particular time and V signifies weight of the output.

The above equations of RNN form the foundation for the transfer functions used to process and interpret data in LSTM. The RNN is basically a mathematical summation process whereas LSTM is more regulated due to its gated structure. Irrespective of RNN, LSTM not only has a feed-forward network to train the network but also has a back-propagation property to be used in the algorithm. A common LSTM involves a cell, which is used to memorize values of the samples at arbitrary time steps, and three gates viz. input, output and forget gate which serves to regulate information with respect to the input.
The sigmoid activation function is embedded to filter the results of all the three gates. A sigmoid activation compresses the input sample values within the range of 0 and 1.Now this step down is used to update or forget data and to simplify data pre-processing. This is mainly because any sample which gets multiplied by 0 or any value tends to zero is taken 0, causing samples to be "forgotten." Similarly, any number in multiplication with 1 is same number therefore the sample is "retained." This way the network can memorize important data.

The following section expatiates about LSTM a modified and gated version of RNN backed with greater prediction accuracy which can solve the vanishing gradient descent problem which are usually due to frequent updates towards a convergent point which can make the results noisy. To avoid this problem LSTM regularization is employed.

1. Forget Gate

First, there is the forget gate. This gate determines which information should be retained or excluded. The position is as indicated in Figure 1(2).

This operation is controlled by a sigmoid function wherein, the information from the preceding hidden state (a_{t-1}) and from current input (x_t) is passed through it for examination. Values come in the range of 0 and 1, such that closer to 0 means to forget and closer to 1 means to retain. The resulting vector is given as:

$$f_t = \sigma\,(W_f \cdot [a_{t-1}, x_t] + b_f)$$

2. Input Gate

Secondly, for updating the cell's state there is an input gate. This operation is divided into two sections which basically consist of an update gate and a tanh function. First, the preceding hidden state (a_{t-1}) and the current applied input (x_t) passes through the update gate, which is basically a sigmoid activation, that decides which values should have an update by squishing all the applied values in range of 0 and 1. Second, the inputs applied to the update gate are also advanced to a tanh function where the values applied are squeezed in the range of -1 to 1, which further outputs a candidate value (c'_t) to update the cell state. Then the tanh output is multiplied with sigmoid function (update) output. The sigmoid counterpart than determines the retained information from tanh output. The structure of this gate is seen in Figure 1(2).
The resulting equations are as follows:

$$i_t = \sigma\,(W_i \cdot [a_{t-1}, x_t] + b_i)$$

$$c'_t = tanh\,(W_c \cdot [a_{t-1}, x_t] + b_c)$$

3. Cell State update

After the above gated manipulations, we now have data suffice to calculate the new cell state information. First, the previous cell state gets multiplied element-wise with the forget vector equation, which basically involves forgetting a piece of previous cell state.

Second, we extract the overall output from input gate operation and add element-wise which changes and updates the cell state to new information that is pertinent for the network for prediction and sequential computation. Hence these two operations provide us with latest cell state. The structure of this gate is seen in Figure 1(2).The equation is as follows:

$$C_t = f_t * C_{t-1} + i_t * c'_t$$

4. Output Gate

The final operation involves the output gate, which decides what the succeeding hidden state should be. This consequent hidden state requires the previous input samples used along with an updated cell state to get a mathematically correct prediction. So here, first, the inputs a_{t-1} and x_t are passed through the output sigmoid function which is essentially a sigmoid activation at the terminal side of the LSTM. Then we pass the modified cell state (C_t), as shown by the equation above, through a tanh function for constriction of values within range of -1 and 1 to obtain a vector compatible with the output of the sigmoid activation, which acts as an adjudicator to determine which values to be carry-forwarded. The output from the sigmoid function (o_t) and tanh functions (C_t) are further multiplied to get the proceeding hidden state.

The structure of this gate is seen in Figure 1(2).. The equation is as follows:

$$o_t = \sigma\,(W_o \cdot [a_{t-1}, x_t] + b_o)$$

$$a_t = o_t * tanh\,(C_t)$$

3. AIS DATASET

Automatic Identification System (AIS) is an autonomous vessel tracking system which is a must on every seafaring vessel. The sensing and measurement devices can be connected and tuned with a broadcast positioning and transceiver system which functions in the VHF mobile maritime band. This system can also be used as bi-directional communication systems for shore based communication. The AIS broadcast static, dynamic, voyage-based and safety-based data. The static data involves: MMSI, IMO number, ship name, call sign, ship's dimensions, ship's category etc. and dynamic data includes ship's coordinates, Universal Time Coordinated (UTC), Course Over Ground (COG), Speed Over Ground (SOG), navigational status and heading.

These parameters of a vessel are simultaneously being plot by a chart-plotter system on an e-display chart. The data is acquired from satellite systems. The major four components which are used in for experiment in this paper are Latitude, Longitude, SOG and COG. These are core dynamic properties which change with respect to time and are used to predict future location of the vessel owing to both the actual speed and the drift due to the environmental conditions.

The real-time data obtained is not consistent many times. There might be any increase in the speed of the vessel, so this might lead to more observational entries per unit time, whereas sometimes there might be some nautical breaks or anchoring or negotiating turns, so these factors might lead to inconsistent data entries leading to great differences in time series plotting. Hence these factors provide us with real time data with respect to the vessel which can be used to model the collision avoidance systems and can also be used steer through the onshore based remote sensing techniques. This data can also be used to predict the future path and course consistency of the vessel owing to the weather conditions. As a result each and every vessel can also be linked within VHF coverage band (20 nautical miles) creating a linked ecosystem with systematic traffic monitoring.

4. FUNDAMENTAL STAGES FOR TRAJECTORY PREDICTION OF ASV USING LSTM

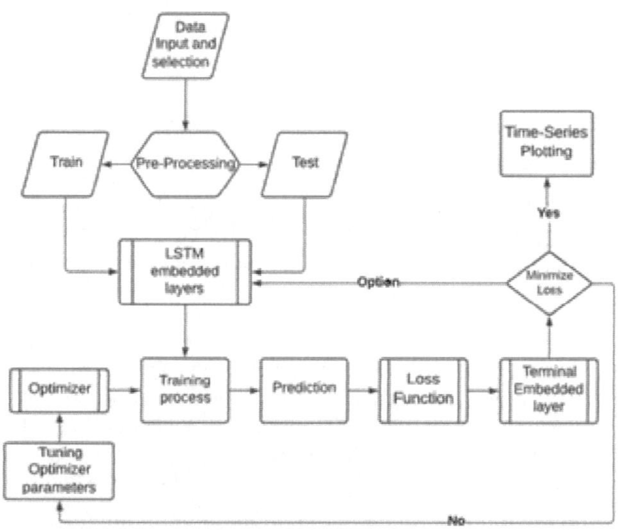

Figure 2: Experimentation process flow

4.1 BASIC PREDICTION METHODOLOGY

The paper posited uses a prediction analysis treatment for the subsequent coordinate points based on the externally supplied ingredient known as the AIS data set. This data set is used to extract the dynamic characteristics namely Latitude, Longitude, SOG, COG, with respect to static characteristic which is the Maritime Mobile Service Identity (MMSI) number.

The idea is taking current position along with the historical position and passing to the network in such a probabilistic range that the transfer functions predefined within the gates in the cell which can then be easily computed and compared to derive results, and this data can be further passed to a suitable function to convert the resulting probabilities in the form of inputs provided earlier at the initiation stage of the cell. In this way real time data is passed through LSTM cells which are embedded in the layers to provide certain level of prediction accuracy which can be taken as a basic threshold to manipulate the hyper-parameters of the network overall to lower digressions from the original data.

In order to express the process in a minimalistic way consider the overall trajectory as L which is dependent upon path points w.r.t time given by $W(t)$. The equation binding these quantities is given as $L=[W(t_1), W(t_2), W(t_3)\ldots\ldots, W(t_n)]$. Here $W(t_n) = [x,y,s,c]$, where x is the latitude and y is the longitude of ship position, s is the SOG and c is the COG.

If the trajectory at $t+1$ is to be predicted where it is a function of n data points which are given as inputs acting as historical positions, the output is given as:

$$W(t+1) = g([W(t+1-n), \ldots\ldots\ldots\ldots, W(t-2), W(t-1), W(t)])$$

4.2 PRE-PROCESSING

The AIS data according to requirements modified such that it can be fed into the LSTM network in shape desired. The stages of pre-processing are:

4.2 (a) Outlier filtering

The AIS data consists of latitude, longitude, SOG and COG. These are the major dynamic characteristics which we use as input, but AIS dataset consists of vessel deflection, angle and track. These data majorly influences the overall position of the vessel.
This kind of data is time varying and irregular. The data might be having track to point drift, which accounts for high deviation within small time gaps, or missing data points which might influence the overall prediction.
Such minute readings which account for errors in the loss function are called as outliers and these must be eliminated.

The method employed here is a standard deviation, wherein the data is randomized and the overall mean and spread (standard deviation) is measured. The anomaly filter is specified for 3 standard deviations. Then the upper and lower limits are set by comparing the mean and anomaly.

Hence, if any data point that exceeds 3 times the standard deviation, then those points are intended as outliers.

4.2 (b) Normalization and reshaping

Normalization is a cardinal feature that includes scaling the input data within a comparable range. The data obtained has many feature vectors which have great difference in magnitude, which might neglect some of the smaller data points. Hence each of the data point gets scaled in a comparable range and normalizing also removes any geometrical biases towards the input features.

The Min-Max normalization method is employed in this paper, given as:

$$y' = \frac{y - min}{max - min}$$

Here, y' is normalized data, y is input data, max and min are the minimal and maximum values of the input data respectively. Once the data passes through this equation the input data magnitude is scaled in the range of [0,1]. The normalized data obtained is reshaped next, by fitting it in a single scalable column to make it individually uni-variate for each time dependent quantity passed as input.

4.2 (c) Bifurcation

The filtered data is further divided into train set, to train the model and yield statistical results, and test set which serves as a procurator for testing the trained model on new and random data.

4.3 NEURAL NETWORK SPECIFICATIONS

4.3 (a) Number of Time-steps

The ANNs usually use vector (rank one tensor) inputs for the modelling of the data, whereas the input fed into LSTM network includes a 3D input. This is because of the sequential nature of LSTM for memorization and comparison of the data requires input in the form of a 3D matrix in the with respect to the three vectors which are Batch size, Time-steps, Feature(size).The basic processing input for LSTM network has input in form of rank three tensor.

Basic RNN uses sidelong accretion whereas there is a selective memorization and cell status update when LSTM is used. LSTM requires certain amount of "look-back" for the network to make a contextual prediction, this interval of look-back is called time-step. It basically means how many past points we have to assert to upgrade present data for prediction.
In experimental analysis of this paper, we have chosen to use 5-7 time steps from the historical positional data. Different values will have a varied amount of impact on the loss encountered during training and also on the architectural arrangement of the network.

4.3 (b) Determination of Number of nodes in the Hidden Layer

Once the above hyper-parameter is declared, another more substantial role is played by the number of nodes in the hidden layer of the network. Hidden layers fill the gap between input and output layer, wherein the function of each node applies specified weights from the weight matrix and administers them through the output activation function.

Each hidden layer has a number of nodes embedded in them. These nodes correlate with the hidden input and output while computing the model, hence playing the role of neurons. The number of neurons is a significant tuning parameter, such that if this number is very small, model won't have enough learning ability to converge at a single result whereas if this number is significantly large then the computation may take a great amount of time hence damping the learning ability of network. Furthermore, the network might not be able to reach a global minima and might end up stuck in local minima giving us impertinent results.

Practical way to find optimum range of these parameters is by repeated experimentation. This method can be easily implemented for small-input linear classifiers but for large-input non-linear relationships this trial-and-error method can be time consuming and result digressing. This problem is omnipresent because real-time AIS data is erratic. In order to make a more generalized network which can reduce this problem, this paper proposes augmentation of formulations derived in[7] with LSTM hyper-parameter tuning. Reference [7] posits that in two-hidden-layer network having e output nodes, the number of hidden nodes suffice to memorize X samples with relatively small loss for,

First hidden layer:

$$\sqrt{(e+2)X} + 2\sqrt{\frac{X}{e+2}}$$

Second hidden layer:

$$e\sqrt{\frac{X}{e+2}}$$

Here feed-forward networks amend the intermittent weights for producing output within an optimal range to match with predefined values, by stepping-down the error.

Further tuning of the layers is required for different data inputs, in case of supervised learning.

4.4 OPTIMIZER

Learning rate defines the amount of model change with each of specified step of training process. It is necessary to reduce the learning rate with each epoch, because as training increases, learning rate also increases with respect to it leading to a higher kinetic energy of the function i.e. parameter vector moves chaotically and may settle in a sub-optimal solution. To prevent this Adaptive moment estimation (Adam) optimizer is used which decays specified learning rate by some amount in each epoch hence preventing the losses. It is an algorithm based on gradient descent required to optimize random seed function. This parameter is tuned by supervised learning when using big and highly stochastic data.

The choice of adam optimizer is justified by its ability to regulate learning rate and configure the steady-state decay per epoch in such a way that output leads to near-accurate predictions in almost any kind of data fed-in hence eliminating the need for learning rate scheduling. Operation of adam can be controlled by four basic parameters viz. α, which is the learning rate; β_1 and β_2, which are the exponential decays for 1st and 2nd moment estimates; and division constant $\mathcal{E}$.

4.5 LOSS FUNCTION

A loss function maps down intuitively a set of parameter values for network on a scalar value that depicts how accurately those parameters have attained the task the network had intended to perform. The loss function used in this paper is Mean Squared Error (MSE), which is widely used loss function. It is given as a summation of square of the difference between the observed and predicted value.

Equation is as follows:

$$MSE = \frac{1}{n}\sum_{i=1}^{n}(y_j - x_j)$$

Where n is number of values, y_j is the observed value and x_j is the predicted value.

5. EXPERIMENTAL APPROACH AND RESULTS

5.1 ENVIRONMENT USED

The hardware set-up used: CPU specification is Intel core i7 micro-processor, memory used is 8GB, the experimental platform is established upon framework of Keras of Tensorflow version 2.3.0, and programming language used is Python 3.7.

5.2 EXPERIMENTAL APPROACH

The methodology follows up as, the AIS data-set of the year 2019 is obtained. The data has different groups of research vessels, such that each vessel has a data timeline of 4-5 hours of offshore coordinates as input. The time-stamp suggests inconsistent data input. The time-stamp is bifurcated into time in hours, minutes and seconds. To make the data entries more frequency dependent and less particular time dependent, the overall time in seconds is considered. A particular category of vessel is picked-up based on MMSI number and subsequently other dynamic characteristics are grouped into a data-frame.

This data is further filtered for outliers and normalized. The normalization is brought into effect by a MinMax scaler of Scikit-learn library. This is followed by reshaping and dividing the data into train and test sets with a division of 80% and 20% respectively. This processed data is passed on to LSTM layers.

The common parameters shared by each univariate data are initialized. The batch size and time-steps and features initialized initially are 1, 7 and 1 respectively. The preliminary activation functions used here are the sigmoid and tanh functions, because the LSTM being a memory block requires a decision to forget or retain data which is given by "0" or "1". The loss function used is MSE for the training phase. In this paper, in order get a salubrious fit of the model the Root Mean Squared Error (RMSE) for train and test prediction is also computed, which assigns large weights to large errors that can bring efficient updates overall. The optimizer availed here is the Adam optimizer instantiated with a learning rate of 0.001 and default values of β_1 and β_2 being 0.9 and 0.999 and $\mathcal{E}$ being 1e-08.

In order to prevent over-fitting, Random module of python is used to generate pseudo-random numbers, also, optionally, we have added, in certain experimental iterations, a dropout layer which randomly drops certain connections intermittent to the hidden layers. Further in order to fine tune the network for accurate predictions the layers and hyper-parameters mentioned above can be experimented.

This paper proposes an initial hyper-parameter tuning based on [7], wherein the layers are approximated in such a way that the optimal range or even accurate values can be obtained by supervised learning.

Finally the prediction plots are juxtaposed with the original data and plotted.

5.3 TUNING BY BIDIRECTIONAL LSTM (Bi-LSTM)

In order to prime network used in our experiment and to achieve pin-point accuracy without much computational experimentation Bi-LSTM layer is also added, which can effectively increase the robustness of the network against any kind of data fed into it.

The advantage of using Bi-LSTM is to appear at a selected sequence each from front-to-back in addition to from back-to-front and during this way, the network creates a context for each character within the text that depends on both its past as well as its future.

Bi-LSTM looks accurately the same as its simplex counterpart LSTM. The distinction is that the network isn't just connected to the past, but also to the future.

5.4 RESULTS

The experimentation was carried out on another 23 different types vessel trajectories, out of which three are depicted in Table 1 and Table 2, to test the effectiveness of the model in lines with reference [7], and it was concluded that the protean hyper-parameter refurbishing process can effectively be approximated in an optimal range without any convoluted permutations within hyper-parameters.

The results are represented as following:

5.4 (a) Individual Feature Prediction Results

Here we take the vessel with MMSI number: 403638000 for our experiment having an average engine speed range of 1.5-3 knots, then we divide the dynamic data in train and test sets with 80% and 20% split respectively. After realizing the approach explained above, the plot of latitude, longitude, SOG and COG are depicted below. It can be clearly seen from the plots below that the positional coordinates, latitude and longitude, almost coincide with the original data from AIS.

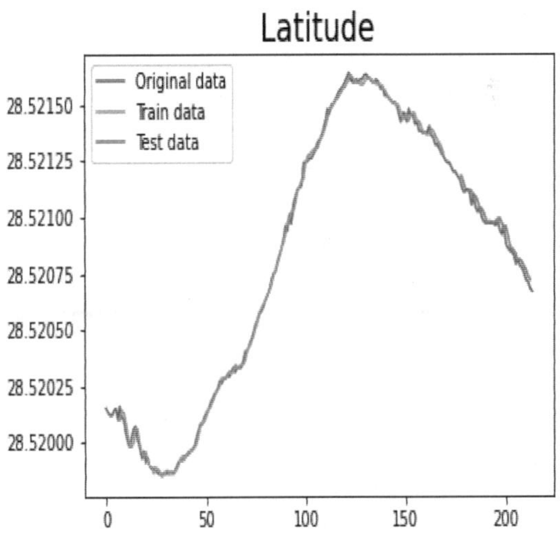

Figure 3: Latitude time-series prediction plot

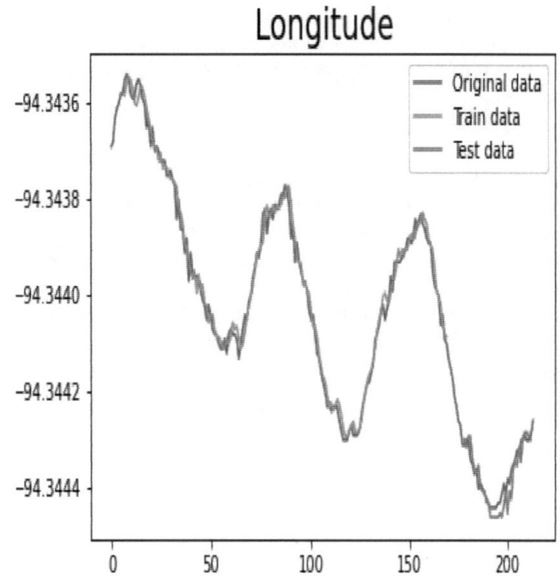

Figure 4: Longitude time-series prediction plot

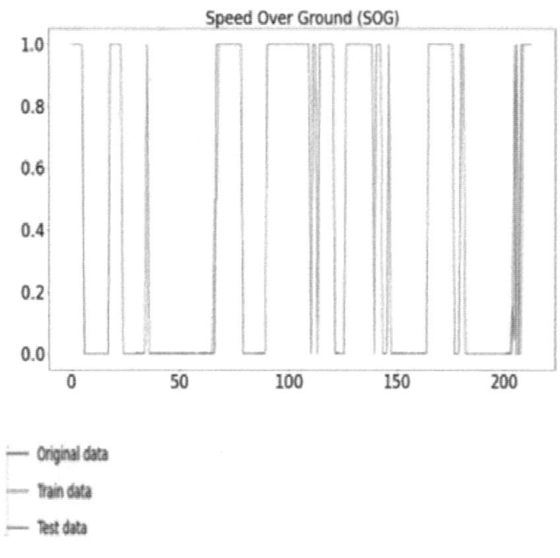

Figure 5: SOG time-series prediction plot

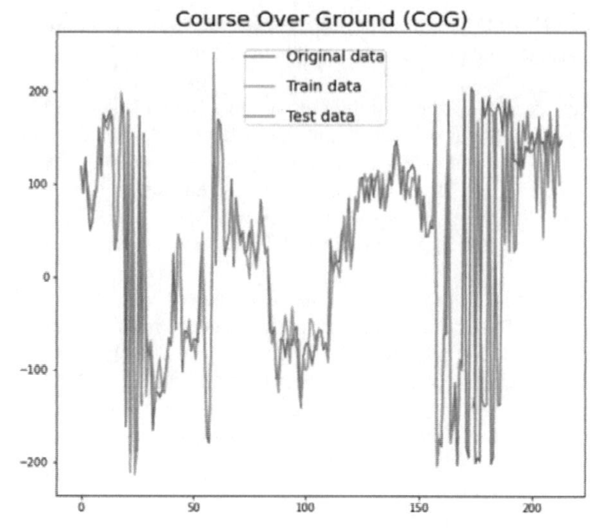

Figure 6: COG time-series prediction plot

5.4 (b) Future Prediction Application Based on Positional Coordinates

The experimental results portrayed above show the approximated trajectory prediction results. Now in lieu of bad weather conditions or due to any offshore signal cut-off or due to any piracy concerns there might be data attenuation from the ASV's transponders, now in such situations the model designed in this paper based on LSTM can even estimate an approximate trajectory for next 30 minutes interval which can provide time window for effective rescue of the ASV with almost a pin-point location accuracy.

In order to actually test the prediction results independently the Figure 7 and Figure 9 below indicate the independent train and test results. These independent plots clearly posit proofs for reliable prediction for next 30 minutes.

As a result the plots below also indicate the original data plus positional prediction data for next 30 minutes combined in a single continuous plot to prevent any aberrations in the readings.

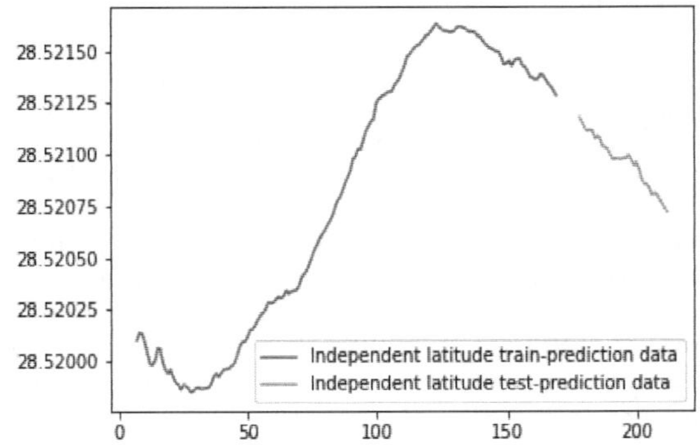

Figure 7: Independent plot (Latitude)

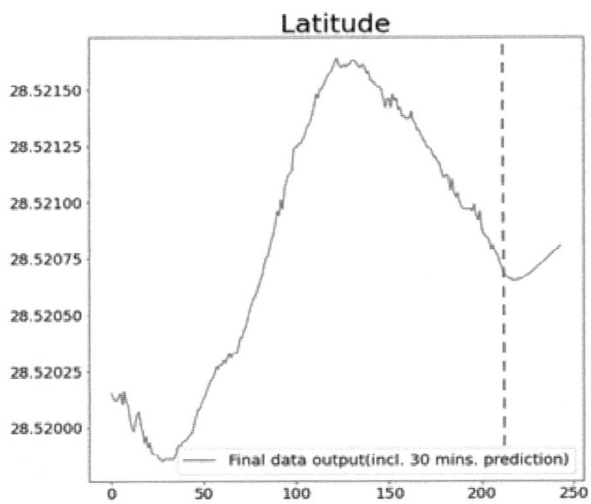

Figure 8: Data output for overall latitude prediction

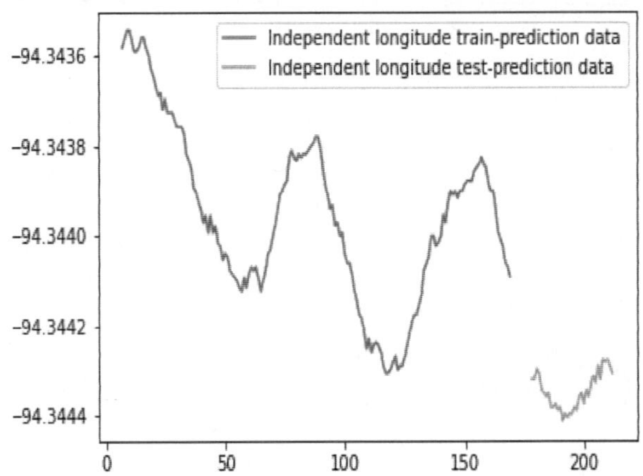

Figure 9: Independent plot (Longitude)

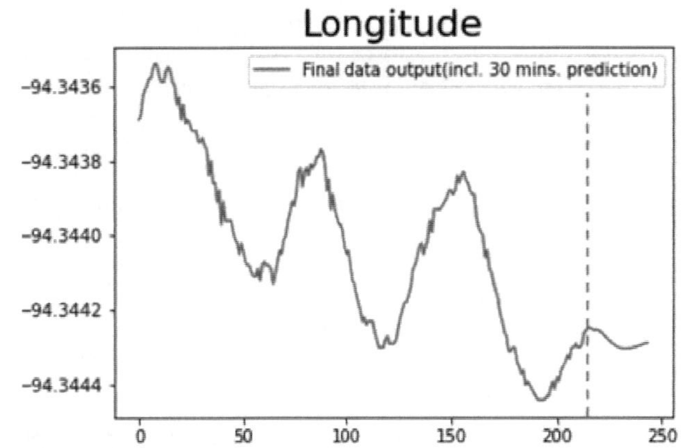

Figure 10: Data output for overall longitude prediction

The path beyond dotted line is the prediction for 30 minutes. The trend of the path plotted is based on compendious effect of preceding experimentation on time series data with optimization of hyper-parameters.

From the plots depicted above, it can be concluded that LSTM neural network can effectively optimize the learning ability and can memorize the input data to provide us an undeviating path depicting the future position, here taken as 30 minutes, in a time series.

5.4 (c) Losses

The plots in our experiment have some losses overall which indicate the prediction error in the neural network. The losses are used to calculate gradients per epoch to update weights and hence optimizing the fit.

In our experiment, the overall calculated error for training and test prediction values is expressed in terms of MSE. The results of MSE are depicted in Table 1 below. To further check the reliability of our model, RMSE loss of the training prediction and test prediction results is separately computed and shown in Table 2 below. The error results are depicted as follows:

Table 1: Results of MSE

		Mean Squared Error (MSE)			
MMSI	Input	Latitude	Longitude	SOG	COG
403638000	214	2.5718e -04	7.8810e-04	3.8010e-03	3.030 0e-02
366978710	430	1.4001e -03	9.5997e-04	7.3061e-05	1.511 1e-05
316031266	398	5.8000e -03	2.2100e-03	1.8002e-03	2.683 4e-04
Average MSE error		2.48e-03	1.32e-03	1.89e-03	1.01e -02

The average MSE is computed, as shown in Table 1, after individual results of latitude, longitude, SOG and COG are deciphered from the network statistics.

The RMSE loss is beneficial for measuring the real-time error directly, because of additional square root which converts input to measurement units.

The LSTM model used here has the capability to pick up characteristics of the input, which is then used to strengthen the memory ability.

Table 2: Results of RMSE

		Root Mean Squared Error (RMSE)			
MMSI	Input	Latitu-de	Longitu-de	SOG	COG
403638000	214	Train: 2.24e-05 Test: 3.54e-05	Train: 2.36e-05 Test: 2.74e-05	Train: 5.98e-02 Test: 5.66e-01	Train: 1.68e-01 Test: 9.63e-01
366978710	430	Train: 4.93e-05 Test: 1.39e-05	Train: 2.58e-05 Test: 1.12e-05	Train: 2.631e-02 Test: 5.86e-02	Train: 3.2e-03 Test: 1.024
316031266	398	Train: 4.71e-06 Test: 1.20e-05	Train: 2.90e-06 Test: 1.13e-05	Train: 8.25e-03 Test: 2.01e-02	Train: 1.54e-02 Test: 2.97e-01

The LSTM model used here has the capability to pick up the characteristics of the input, which is then used to strengthen the memory ability.

The conclusion that can be derived from the above two tables is that no matter how many data points or variations are provided to the LSTM model, it can obtain almost accurate predictions.

6. CONCLUSIONS

Pertaining to the attributes of AIS data this paper presents a LSTM recurrent neural network based model for effectively predicting the future position of an ASV. Latitude, Longitude, SOG and COG are used as eigenvalues for processing of the data.

The posited network is trained by comparing and estimating the error generated between the predicted value and given value and hence it can be concluded that the model's accuracy can be greatly enhanced by data pre-processing and hence LSTM model can memorize and predict the vessel's future coordinates. To depict the application of this approach, LSTM network is calibrated to predict positional data points 30 minutes ahead. To depict the losses incurred during the training of the model, error values based on MSE and RMSE are tabulated for 3 samples.

From the results it can be seen that further tuning of hyper-parameters is required to reach pin-point accuracy in prediction data. Advancements in LSTM network can also be made by tuning hyper-parameters automatically.

7. ACKNOWLEDGEMENTS

The experiment in this paper is carried out on data-set taken from www.marinecadastre.gov.

8. REFERENCES

1. Xilin Liu., 'A new prediction model for ship navigation GPS positioning Trajectory', *Journal of jiangnan university: natural science*, 2014.

2. Qiao SJ, Jin K, Han N, Tang CJ, Gesangduoji, GL., 'Trajectory prediction algorithm based on Gaussian mixture model', *Journal of software*, 2015.

3. X Pecher P, Hunter M, Fujimoto R., 'Data-Driven Vehicle Trajectory Prediction', *SIGSIM-PADS '16: SIGSIM Principles of Advanced Discrete Simulation,* 2016.

4. X Ying J C, Lee W C, Weng T C, 'Semantic trajectory mining for location prediction', *ACM Sigspatial International Conference on Advances in Geographic Information Systems*, 2011.

5. X Gan S, Liang S, Li K, 'Ship trajectory prediction for intelligent traffic management using clustering and ANN', *Ukacc International Conference on Control. IEEE*, 2016.

6. X Graves A, Mohamed A R, Hinton G, 'Speech Recognition with Deep Recurrent Neural Networks', *International Conference on Acoustics, Speech And Signal Processing. IEEE*, 2013.

7. X Guang-Bin Huang, 'Learning capability and storage capacity of two-hidden-layer feedforward networks', *IEEE Transactions on Neural Networks*, 2003.

9. AUTHORS BIOGRAPHY

Dhyey Rajani is currently pursuing his B.Tech in Mechanical Engineering at Dwarkadas Jivanlal Sanghvi College Of Engineering (DJSCE). His previous paper has been published in ICIMA 2020 conference under the title, "A review of Carbon-Nanotubes as Novel Drug Carriers in Cancer Therapy".

DESIGN OF AN UNMANNED SURVEILLANCE BOAT

A.J. Najeefullah, Tecon SRL Engineering Consultancy, UAE

SUMMARY

An unmanned surveillance boat was designed to give solution for coast guard to improve coastal security, help to minimize manpower cost through remote operation and better performance at any environmental condition. This study also proves to maneuver the craft in tight canals with effective management from the shore and minimize construction cost.This paper summarizes the design phases such as selection of hull type, evaluation of principle dimension by considering several factors and scantling calculation. Further, stability, resistance, powering and seakeeping performances are analyzed using Maxsurf software (Bentley Product).

NOMENCLATURE

β	Deadrise Angle (°)
ρ	Density of water (kg m^{-3})
Δ_{WL}	Design Displaced weight (kg)
∇	Design Displaced volume (m^3)
Δ_{Full}	Full Displaced weight (kg)
A_P	Waterplane area (m^2)
B_{WL}	Breadth at design waterline (m)
C_B	Block Coefficient
C_P	Prismatic Coefficient
D	Depth (m)
DLR	Displacement Length ratio
F_r	Length Froude number
F_{rb}	Beam Froude number
F_{rv}	Volumetric Froude number
g	acceleration due to gravity (m s^{-2})
GM	Metacentric Height (m)
GZ	Righting Lever (m)
L	Design waterline length (m)
LOA	Length overall (m)
T	Design draft (m)
V	Maximum Speed (knots)

1. INTRODUCTION

The increasing demand in marine safety, national security, marine search operation and environmental protection, reveal that a greater number of boats with more features are required. With this, the Autonomous unmanned surveillance boat is introduced to fulfill the requirements efficiently and is now a trend in the market.

In recent years the operation of high-speed planing craft for military and commercial uses has been increased. Therefore, the planing hull has been chosen and investigated with several factors that influence high speed. Selection of materials and structure has led to the development of stronger hull, yet keeping the craft light in weight.

The basic principle behind a planing hull is that of the generation of sufficient hydrodynamic lift to the bottom of the hull to support the vessel's weight. The lift force is created when the hull bottom surface is faced with the incoming water at correct angle of attack. Therefore, the effect of hydrodynamic forces is higher than the hydrostatic forces in planing hull. This study examines the final particulars carefully using various design parameters, large angle stability analysis in both intact and damaged conditions, resistance and powering estimations and seakeeping performance of the designed craft.

2. KEY FEATURES

Remotely operated craft, requires additional equipment for monitoring, than manned craft. Advanced robotics and surveillance equipment such as gimbal stabilized camera with an uninterrupted 360-degree view, and Long-range Acoustic Device (LRAD) that communicate with the Shore Control Centre (SCC) in real time are installed to satisfy the standard requirements. It also has a drone which is operated from SCC. An intermediate receiver is installed in the craft to get a better access to the drone, thereby resulting in an effective surveillance.

System that detect and indicate any upcoming technical problems long before the equipment fails are fitted for maintenance schedule. Automatic fire detectors are mounted and CO_2 fire extinguisher have been used as an advantage to the absence of people onboard. Effective sensors are used across the outer hull of the craft to avoid any collision with other vessels especially in heavy traffic regions.

Good instrumentation with visible light and better communication bandwidth are installed for operation of the craft. It also has high capacity data link for direct transfer to HDTV, sensor and guidance signals. Cyber security is a critical issue for the unmanned vessels; therefore, all equipment is encrypted to safeguard it from any illegal operation of the craft.

3. EVALUATION OF MAIN DIMENSIONS

The unmanned craft parameters need to be screened in order to optimize the hull efficiency, length, breadth, block coefficient and longitudinal center of floatation, among others.

Following factors are considered to improve the performance of the craft [5].

- Low Length beam ratio, i.e. ($L/B_{WL} < 3.5$), shows that it requires significantly higher power to weight ratio at pre-planing state, but once the hull is in full planing mode, the power to weight ratio is of same order with other hulls of higher length beam ratio. Wider transom helps to reduce input power as the craft lift easily during planing stage. It is generally agreed that increasing the size of the craft will produce a better design in terms of performance and comfort; on the other hand, the craft might be more difficult to maneuver.
- The craft is light in weight, when fully loaded displacement (Δ_{Full}) is lesser than $(0.13LB_{WL})^{1.5}$, displacement length ratio ($DLR=\nabla/(L/30.5)^3$) is between 100 to 200 and size weight ratio should be high ($A_P/\nabla^{2/3}$), which indicates that the hull plains easily with less input power.
- The craft is in high speed, when maximum speed (V) in knots is equal to or exceeding ($7.16\Delta_{WL}0.1667$) and speed ratio ($V/\sqrt{L}$) is greater than 3.0.
- In experimental point of view, the craft is in planing condition when $F_r > 1.0$ or $F_{rb} > 3.0$ or $F_{rv} > 4.0$, where $F_r= V/\sqrt{gL}$. As the waterline length is variable, $F_{rb}= V/\sqrt{gB_{WL}}$ and $F_{rv}= V/\sqrt{g\nabla^{1/3}}$ are introduced.
- Deadrise is the angle a hull bottom makes with the horizontal plane. The correct amount of deadrise gives the boat directional stability, a softer ride and reduces wetted surface drag as the boat rises on a plane. In this craft, we have monohedral hull, i.e. constant deadrise from midship to transom of 10°. Whereas in forward, the deadrise is 22°. Larger the deadrise, more the power consumption with smoother ride even in extreme environmental condition, thereby preventing any blur in capturing the real time images.

Taking above factors into account three-dimensional hull form was developed using Rhino 3D and imported in Maxsurf Modeler. Design process involves the use of knowledge and visualization of the hull shape to achieve the correct design parameters. The derived main dimensions and design parameters are listed in Table 1.

CRAFT'S MAIN DIMENSIONS		**DESIGN PARAMETERS**		
LOA (m)	4.90	L/B_{WL}	3.0	< 3.50
L (m)	4.50	Δ_{Full} (kg)	776	< 810
B_{WL} (m)	1.50	*DLR*	182	100<DLR<200
D (m)	0.77	$A_P/\nabla^{2/3}$	6.8	High
T (m)	0.24	$V/\sqrt{L}$	5.8	> 3.00
Δ_{WL} (kg)	600	*V* (knots)	24.0	≥ 6.30
∇ (m^3)	0.59	F_r	1.9	> 1.0
A_P (m^2)	4.75	F_{rb}	3.2	> 3.0
C_B	0.36	F_{rv}	4.3	> 4.0
C_P	0.65	β (°)	22.0	< 25
V (knots)	24.0			

Table 1: Principle Dimensions and Design Parameters

4. SCANTLING

For planing hull, weight is a concern, as each additional weight will reduce the hydrodynamic forces and increases the input power. Therefore, it is necessary to determine the required scantling arrangement for the craft carefully. Scantling dimensions are derived using scantling number which is dimensionless and then examined with the allowable section modulus and pressure [1] to finalize the required scantling. The Scantling number is determined based on length, beam and depth of the hull [3]. The final structural components are evaluated and listed in Table 2.

4.1 MATERIAL SELECTION

Aluminum is chosen as it is lighter than steel with the same strength. This will lower the center of gravity of the craft, making it more stable and thus more seaworthy. Lesser the weight, faster the craft with the same input power. The main advantage of the aluminum is it does not rust easily and has low maintenance cost. As it is corrosion resistant, additional plate thickness for corrosion allowance is not required. This in turn reduces the additional craft's weight. According to DNV [1], VL5083 grade is used for plates and for the profiles VL6061.

4.2 PLATING

The bottom plate of the craft is heavier, to make it more rigid, sustain damage on grounding, reduce vibration and increase its life span thereby making it efficient to move in narrow channels. However, this thicker bottom is needed only at the aft and shall be reduced at forward of about 20 to 25% to reduce weight, yet same thickness has been taken into account to be heavy during slamming at forward. The chine plating has same thickness as bottom plate.

The upper third of the side plate below sheer can be reduced in thickness to that of the deck plate for weight reduction. In this craft, the fully loaded waterline is just above the chine line, and the pressure on the side plate is less, therefore the side plate and deck plate thickness are considered the same. Plating thickness is derived using empirical formula [3] is listed in Table 2.

4.3 BULKHEAD

The forepeak or collision bulkhead is positioned as per DNV standards [1]. All other watertight bulkheads are located at each end of the machinery space. Special bulkhead is provided at aft to protect battery system in case the transom is damaged and also increase strength against vibration. The transom plate thickness has followed same thickness of the bottom plating to reduce vibration and noise created by propulsive motor attached to the transom plate. The other watertight and non-watertight bulkheads are kept with similar thickness as the side plate, considering flooding of water head. Evaluated bulkhead plating thickness is listed in Table 2.

4.4 STIFFENERS

The stiffeners spacing are chosen in such a way to ensure adequate panel stiffness and increase its strength. In this craft, same conservative spacing has been followed at all panels in order to make construction easy. 80% of the spacing is reduced only at bottom longitudinal at forward, to provide more strength during slamming. Both transom and bottom stiffeners have the same spacing with connecting brackets.

Unequal angle bars are chosen to provide enough strength with less weight. The watertight bulkhead stiffeners are equal in size to side shell by taking into account the water head pressure incase of any damage. Centerline vertical keel is a solid section which gives high strength during slamming with 40x5.18mm in thickness. The size of the stiffeners in transom and collision bulkhead are provided similar to bottom stiffeners. Summary of stiffeners are shown in Figure 1.

4.5 DESIGN PRINCIPLE LOADS

The strength of the craft depends on the internal and external pressures such as slamming on bottom, pitch slamming and sea pressure. However, the design of sea pressures is assumed to be acting on the craft's outer panels at fully loaded draught. All these pressures are computed using given empirical formula in DNV standards [1]. Using these design forces, the required thickness is determined and shown below in Table 2.

With these scantling arrangements, the light craft weight of the aluminum vessel is approximately 225kg.

	DNV [1]	**Dave Gerr [3]**	**Final**
Bottom Plating (mm)	4.25	4.18	**4.30**
Side Plating (mm)	2.73	2.66	**2.75**
Transom & Collision Bulkhead (mm)	3.34	2.66	**4.30**
Other Bulkhead (mm)	3.12	2.66	**2.75***
Deck Plating (mm)	3.17	2.66	**2.75***
Bottom Stiffeners (mm)	0.9	0.6	**1.10**
Side Stiffeners (mm)	0.23	0.2	**0.30**
* Stiffeners are taken into account, therefore (10% to 15%) reduction in plate thickness			

Table 2: Scantling Selection

4.6 HULL GIRDER LOADS

Longitudinal strength analysis is performed for the high-speed craft by determining required hull section modulus. In High speed craft, both crest and hollow landing conditions, that arise due to bottom slamming around midship area and fore/aft areas respectively is critical. Longitudinal bending moment is also determined in still water and wave condition. In order to obtain the required hull section modulus [1], maximum bending moment is considered from all the four conditions resulting in 2.69cm^2 m.

The actual section modulus is 27.0cm^{2}m, computed using the midship section as shown in Figure 1, is greater than the required modulus. This proves that the provided scantling arrangement is more effective against all loads.

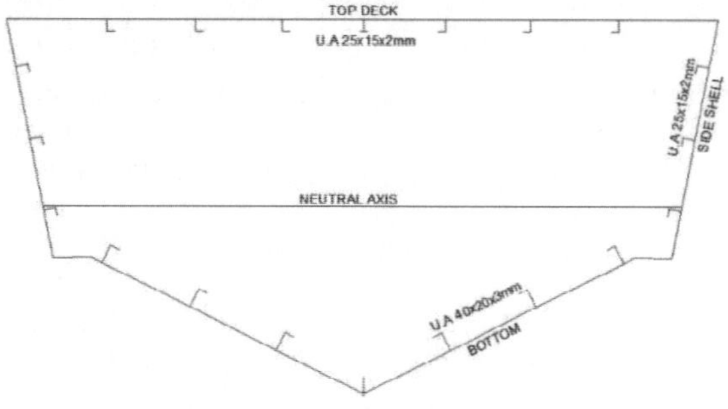

Figure 1: Midship Section

5. RESISTANCE AND POWERING

5.1 RESISTANCE

Hydrodynamic analysis is crucial for planing hull craft. There are several methods to evaluate total resistance, where the best prediction method was used from Savitsky 1964 [4]. The resistance and powering analysis are performed using Maxsurf software [6] and the final results are reported in Figure 2, which illustrates that, when speed increases, the total resistance which includes bare hull, residuary, and friction resistance are reduced as a result of the reduced wetted surface area. This shows that the planing hull is raised and skims onto the surface of the water.

The initial trim angle is 1° by aft at an equilibrium condition of fully loaded case. When the craft is in operation, this trim angle is increased upto 6.72^o and gradually reduces as speed increases. This shows that the running trim angle and resistance are directly proportional, and as they increase, the wetted surface is also increased.

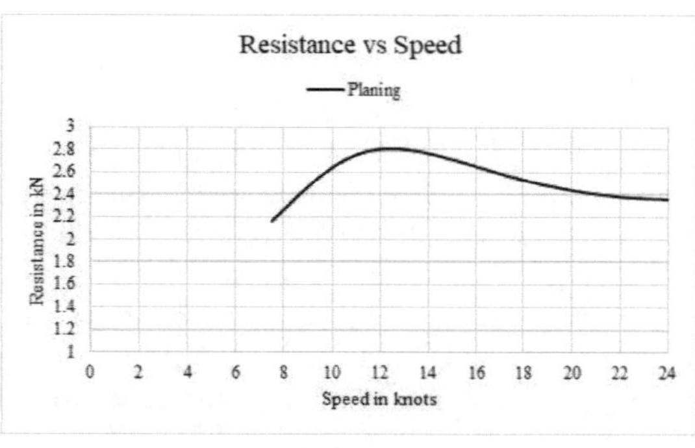

Figure 2: Resistance VS Speed

5.2 POWERING

Power prediction is usually based on resistance computation and an estimation of overall efficiency. The resistance of high-speed vessels is primarily a function of the vessel's displacement, wetted length, breadth and speed.

To reach a certain speed under power with a limited power source the displacement-length ratio is of vital importance. As this ratio is high, the wetted surface area is more, which increases the resistance and requires more power to reach the desired speed. Therefore, this craft has been analyzed with this ratio to fulfil the requirement as shown in the Table 1.

5.3 SELECTION OF MOTOR AND BATTERY

The selection of motor and battery has been done carefully as it affects speed and weight of the craft. As the main aim of the study is to keep the craft light in weight to plain the hull easily. Selection of motor influences the input power, total weight, and LCG position of the craft.

Electric motor has been used as it is powerful, safe and easy to maintain. This motor does not need any submerged appendages and increases the overall efficiency as the restrictions on propeller diameter, normally imposed on a propeller behind the hull, is relaxed.

In order to achieve full speed of 24 knots, 16.0kW propulsive power is required. "Deep blue 25RL" the outboard electric motor has been chosen whose input and propulsive power are 27.6kW and 16.2kW respectively, whereas total gross weight of the motor is 140kg.

Battery power resource has been used for the entire operation of the craft. "BMW 3" battery is chosen, as it has capacity of 40kWh, of which 10kWh for emergency purposes and the craft runs at full throttle for approximately 1.5 hours continuously with the remaining 30kWh. The total gross weight of the "BMW 3" battery is 280kg. The battery is placed in such a way that the length of the wirings are reduced to lower the weight of the craft.

6. STABILITY

Stability represents perhaps the most important aspect of a design since the properties of a craft in these respects reflect its ability to carry the required weight and to withstand the heeling moment. Stability checks have been performed based on International standards [2]. The large angle analysis is performed using Maxsurf stability [7] as it is critical, considering both intact and damaged cases along with light craft and fully loaded condition. There are four compartments having the possibilities for damage, which is considered and analyzed.

Dynamic stability due to large centrifugal forces, generated when a craft changes its direction at high speed, creating a large moment about center of gravity, which heels the craft further, is controlled by lowering the center of gravity.

Suction force created due to heeling at the larger curvature of the chine is reduced by keeping the chine line and wetted part relatively straight. However, the suction force created at curvature of forebody, which is inevitable, can be reduced by controlling the running trim angle. The running trim angle in this craft is kept minimum of approximately 3° by aft at high speed, to reduce the creation of suction force at forward.

6.1 LOADCASES

Lighcraft weight is computed from maxsurf for the plates and calculated manually for the stiffeners as per scantling calculation. The machinery and equipment weights are also included in the same. The total light craft weight is 458kg, whose mean draft is 0.195m at amidships and the trim angle of 1.35^{o}. The vertical height of the center of gravity is 0.521m from the keel.

The battery weight and the weight of CO_2 fire extinguishers are considered in fully loaded case, as it is variable. The total fully loaded case weight is 776kg, whose mean draft is 0.266m at amidships and the trim angle of 1°. The vertical height of the center of gravity is 0.435m from the keel, which is approximately at center of the craft's hull.

6.2 LARGE ANGLE STABILITY ANALYSIS

The large angle stability analysis is used to calculate the righting lever and other hydrostatic data at each heel angles by balancing the defined loadcase displacement against the hull buoyancy.

The key output value is GZ (or righting lever), the horizontal distance between the center of gravity and buoyancy. One of the GZ curve graphs has been depicted in Figure 3. The other information is examined from the GZ curve from the selected criteria [2]. The wind pressure has been computed as 295.0Pa conservatively assuming the extreme wind speed of 20m s^{-1}.

Important information to be noted from the Figure 3 are listed below;

- The maximum GZ of 0.256m, occurs at an angle of 50°, which is the maximum heeling moment required to bring the craft back to its position. Beyond this angle, the stability of the craft decreases drastically.
- The point of vanishing stability occurs at 90° where the righting lever is zero, any further heel beyond this angle would result in a condition of negative stability.
- Craft with a smaller GM will have a longer roll period, which puts the craft under risk for larger angle of heel. The initial GM should not be less than the 0.15m as per the standards [2]. From the Figure 3, the GM shall be measured as 0.675m at 57.3^{o}, which is greater than the required.

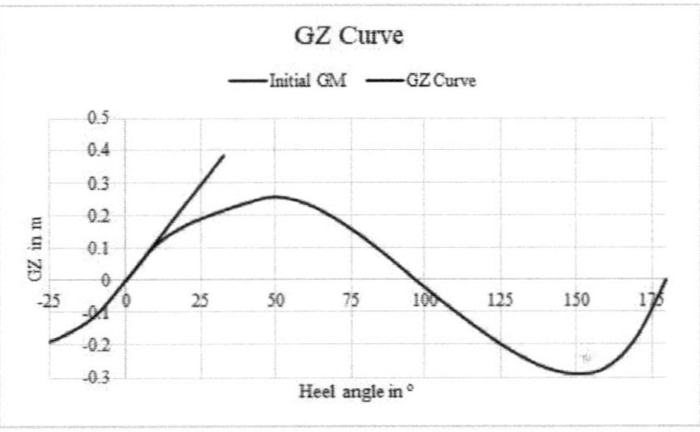

Figure 3: GZ Curve – Intact case -Fully Loaded condition

7. SEAKEEPING

The planing hulls almost inevitably have V-shaped section. The reason for this is the seakeeping qualities of the hull. This reduces the vertical acceleration, thus preventing extremely bumpy rides. Deeper the V, smaller the acceleration.

Linear ship theory [9] is the standard method used to generate transfer function and apply it to the power spectrum of the waves at a particular speed and heading. Several scenarios are created by altering the sea state, speed and heading to understand the craft's performance. The responses in heave, pitch and roll are evaluated and combined to find the craft's dynamic stability at sea. However, this method is too complicated for a project of this size, hence Maxsurf motion software [8] was used, to check the performance and responses of the craft by making the assumption of various scenarios.

Figure 4 shows the craft's transfer functions or RAO. By studying the curves, it is seen that it tends to unity at low frequency, making it simply move up and down along the waves. It is seen that the response approaches to zero at larger frequency, due to the effect of small waves cancelling each other over the length of the craft. The RAO value greater than unity means that the craft's response is greater than the wave amplitude.

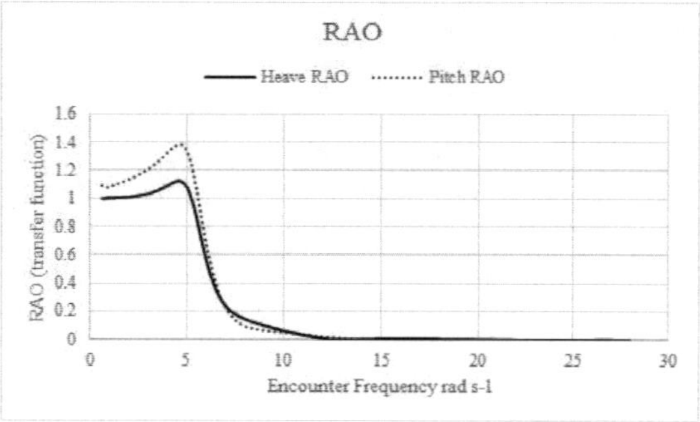

Figure 4: RAO in heave and pitch

8. CONCLUSIONS

The unmanned surveillance craft has been designed by various analysis including resistance, powering, stability and scantling calculation.

The Final main dimensions are evaluated carefully by considering several design parameters to make the craft plain easily and perform better. It has low length beam ratio which makes it possible to pass through narrow channels. Aluminum material is used to reduce the craft's weight as it is of more concern in planing hull. The strength of the material is in accordance with codes and standards [1].

The craft has low resistance force as its wetted surface area reduces gradually as speed increases. Highly rated eco-friendly electric motor and battery are chosen as per the power requirement. It can operate efficiently for about 1.5 hours.

The craft is stable at any angle of heel, both intact and damage cases. It also complies with the rules and regulations [2], in order to sail safely and efficiently throughout its life time. The running trim angle and center of gravity are kept low to reduce the dynamic stability. Furthermore, the craft performs well in seakeeping even at irregular waves.

Manpower cost is reduced as there are no people onboard, and also requires minimum expertise for shore control center. There is no consumption cost as battery is used instead of fuel. Maintenance cost is also reduced, as corrosion resistant material is used, less chances of grounding due to low draft and heavier bottom which lowers the damage due to slamming.

Thus, this high-speed craft has a longer lifespan, is stable and performs well in rough weather conditions, captures high quality images for surveillance and is cost effective.

9. REFERENCES

1. 'Rules for Classification High Speed and light craft', DNVGL-RU-HSLC-Pt3, 2018.

2. 'International Code of Safety for High-Speed Craft', HSC 2000 Code, 2000.

3. DAVE GERR., 'The Elements of Boat Strength for Builders, Designers and Owners', 2000.

4. D. SAVITSKY, 'Hydrodynamic Design of Planing Hulls', *Marine Technology, Volume 1*, 1964.

5. LARS LARSSON and ROLF E ELIASSON, 'Principles of Yacht Design', *Adlard Coles Nautical, Volume 1*, 2000.

6. 'MAXSURF Resistance Program & User Manual', *Bentley Systems, Incorporated*, 2018.

7. 'MAXSURF Stability Program & User Manual', *Bentley Systems, Incorporated*, 2018.

8. 'MAXSURF Motions Program & User Manual', *Bentley Systems, Incorporated*, 2018.

9. EDWARD V. LEVIS, 'Principles of Naval Architecture – Motions in Waves and Controllability', *The Society of Naval Architects and Marine Engineers, Volume III*, 1989.

10. AUTHORS BIOGRAPHY

NAJEEFULLAH holds the current position of Naval Architect and Structural Engineer at Tecon SRL. He is responsible for performing offshore structural analysis using SACS software, preparing mooring plan and analysis. His previous experience includes performing stability calculation for ships during docking. He is professionally a Naval Architect and offshore engineer, and has completed university thesis of "Initial Design of LNG Carrier" successfully.

MULTIVEHICULAR UNDERWATER TRANSPORTATION USING FLEXIBLE CONNECTIONS

F. U. Rehman, E. Anderlini, G. Thomas, University College London, UK
A. Ali, Punjab Group of Colleges, Pakistan

SUMMARY

Underwater transportation using multiple Autonomous Underwater Vehicles (AUVs) could be useful for both the military and commercial sectors. In this paper, a flexible connection transportation system of two Hovering AUVs was analysed. A nonlinear dynamic model was developed using the Langrangian approach in which the payload was taken as a reference. Initially, the significance of underwater buoyancy was elaborated. Subsequently, it was worked out that the payload cannot be made neutrally buoyant or positively buoyant due to the adverse effect on payload stability and motion response. Moreover, it is a difficult and hectic process to get desired swing angles of the flexible links by manual selection of the thrust forces, for motion in all directions. Therefore, a control system was designed first using the PID controller and then the optimal pole placement controller. The PID controller saved a lot in power consumption but compromised on the motion response and vice versa for the pole placement controller.

NOMENCLATURE

L	Langrangian operator
T	Total kinetic energy
U	Total potential energy
q_k	Generalised coordinates
Q_k	Force applied in q_k
O	Centre of body of the system
COG	Centre of gravity
COB	Centre of buoyancy
EFF	Earth-fixed frame
BFF	Body-fixed frame
M_t	Mass of the system
M_1	Mass of HAUV 01
M_2	Mass of HAUV 02
M_p	Mass of payload
l_1	Length of flexible link 01
l_2	Length of flexible link 02
(I_{y_1}, I_{y_2})	Pitch moment of inertia of HAUV01 and HAUV02
(x_p, z_p)	Surge and heave of payload
(λ_1, λ_2)	Swing angles of flexible links
(θ_1, θ_2)	Pitch angles of HAUV01 and HAUV02
$(X_{\ddot{x}}, Z_{\ddot{z}}, M_{\ddot{\theta}})$	Added mass terms of the system
$(X_{\dot{x}}, Z_{\dot{z}}, M_{\dot{\theta}_1}, M_{\dot{\theta}_2})$	Damping terms of the system
D_{t_x}	Total surge damping of the system
D_{t_z}	Total heave damping of the system
D_{t_1}	Translational damping terms of HAUV01
D_{t_2}	Translational damping terms of HAUV02
$(X_{\ddot{x}}, Z_{\ddot{z}}, M_{\ddot{\theta}_1}, M_{\ddot{\theta}_2})$	Added mass term of the system
V_t	Volume of the system
V_1	Volume of HAUV01
V_2	Volume of HAUV02
V_p	Volume of payload
$\boldsymbol{T_a}$	Thrust allocation matrix of the system
$\boldsymbol{T_{a1}}$	Thrust allocation matrix of HAUV01
$\boldsymbol{T_{a2}}$	Thrust allocation matrix of HAUV02
$\boldsymbol{f_{a1}}$	Thrust force vector of HAUV01
$\boldsymbol{f_{a2}}$	Thrust force vector of HAUV02
$\boldsymbol{M(\eta)}$	Mass matrix of the system
$\boldsymbol{D(\eta, \dot{\eta})}$	Damping matrix of the system
$\boldsymbol{g(\eta)}$	Vector of hydrostatic terms
K_P	Proportional gain
K_i	Integral gain
K_d	Differential gain
$\boldsymbol{\eta_d}$	Vector of desired state
$\boldsymbol{\eta}$	Vector of actual state
$\boldsymbol{e}$	Error between $\boldsymbol{\eta_d}$ and $\boldsymbol{\eta}$
$\boldsymbol{\tau_d}$	Desired thrust vector of the system
$\boldsymbol{\tau}$	Actual thrust vector of the system
$\boldsymbol{f_d}$	Desired thrust force vector of the system
$\boldsymbol{f}$	Actual thrust force vector of the system
$\boldsymbol{A}$	System matrix
$\boldsymbol{B}$	Input matrix
$\boldsymbol{C}$	Output matrix
$\boldsymbol{D}$	Feed forward matrix
$\boldsymbol{x}$	State vector of the system
$\boldsymbol{K}$	Pole placement gain matrix
RMSE	Root Mean Square Error
RMS	Root Mean Square

1. INTRODUCTION

The underwater transportation could be used in the military sector for transportation of a payload which is required to be unseen or less exposed to detection. Moreover, it could be employed in the offshore industry for precise underwater installations.

There are generally two ways to connect the autonomous vehicles to a payload i.e. rigid and flexible connections. The transportation system having rigid connections are easy to analyse due to the consideration of the whole system as a single rigid body [1]–[4]. This requires the development of a single dynamic model for the entire system in which the translational parameters are summed together and the effect of rotational parameters are taken about the centre of the combined body i.e. the centre of the payload [5]. Moreover, a centralised control system is sufficient for the combined system to follow the desired trajectory [6]. On the other hand, a flexible connection transportation system depends on the stability of the individual vehicles and their capability to maintain the desired position of the payload and the swing angles of the flexible links [7]–[11].

Remotely Operated Vehicles (ROV) can move in any direction with precision due to over-actuation. However, it can only move a limited distance depending on the length of the cable which is connected for providing power and communication. In this paper, ROV was used autonomously and was called the Hovering Autonomous Underwater Vehicles (HAUV).

Minerva ROV was selected to be used autonomously due to the detailed availability of its model parameters in the literature [12]–[14]. The specification of Minerva HAUV is shown in Table 1. Minerva HAUV is both statically and dynamically stable. It is designed to have its buoyancy only slightly higher than its weight so that lesser power is applied to keep it submerged, meanwhile, provide it with the capability to come to the sea surface in case of an emergency [13]. In the simulations, Minerva HAUV is considered neutrally buoyant to ease the analysis.

Table 1: Specification of Minerva HAUV

Specification	Minerva HAUV
Mass (Kg)	520
Length (m)	1.5
Width (m)	0.82
Height (m)	0.84
Number of thrusters	05 (02x vertical, 02xaxial, 01 transverse)
Force of thruster	Maximum = 340N Operation = 300N

The rest of the paper is organised as follows. Section 2 describes the system for which the analysis is required. In Section 3, a nonlinear dynamic model is developed to represent the actual system in the simulations. In Section 4, several tests are performed to verify the simulation code. Subsequently, tests are performed in Section 6 to select the payload mass for getting an appropriate motion response. The control system is designed in Section 7 and implemented on the system in Section **Error! Reference source not found.**. Finally, concluding remarks are given in Section **Error! Reference source not found.**.

2. SYSTEM DESCRIPTION

Two neutrally buoyant Minerva HAUVs are connected to a cubic payload via flexible links, as shown in Figure 1. The length of the flexible link is 3m as this is practically feasible. The payload is a container of volume $1m^3$ in which the payload of a known weight is filled. The buoyancy force of the payload becomes 10055N. Though it is practically desirable to make the payload slightly positive buoyant so that the whole system comes to the sea surface under emergency. However, it is not acceptable for the flexible connection system as discussed in Section 5.c.

Initially, the payload is at (0m,0m) whereas HAUV01 and HAU02 are at (0m,-3m). Therefore, there is zero swing angles of the flexible links.

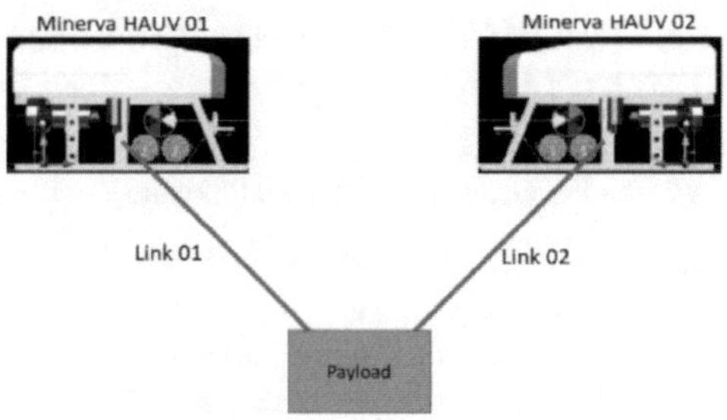

Figure 1: Flexible Connection Transportation System

3. DEVELOPMENT OF THE DYNAMIC MODEL

The dynamic model was developed in the vertical plane using the Lagrangian approach. The payload was taken as a reference and the kinematics of HAUVs were defined accordingly. It was considered that the pitch motion can be controlled independently by each HAUV. Therefore, it was ignored in the dynamic model.

The following assumptions were taken in the development of the dynamic model:

1. The HAUVs and payload are rigid bodies of constant mass.
2. Mass and its distribution on the system do not change during motion.
3. The system is deeply operated so the wave effects are ignored.
4. The flexible links are massless.
5. The flexible links are un-stretchable and cannot have a sag.
6. The interaction with other sea bodies is ignored.
7. The Coriolis forces and moment are ignored.
8. One end of the flexible link is attached to the centre of HAUV and other to the centre of payload.

Two reference frames are generally used i.e. Earth Fixed Frame (EFF) and Body Fixed Frame (BFF). The xy-plane of EFF is tangential and xz-plane is perpendicular to the surface of the earth. They are fixed on earth and do not vary with the motion of the system. On the other hand, BFF has its axes fixed on the system and vary with the motion of the system.

In this analysis, the derivation was done in EFF as the relative motion between the HAUVs and payload occurs in the inertial frame. The pitch motion of HAUVs was ignored as they are uncoupled from the system's dynamics and would be controlled independently by the HAUVs.

The equations of motion of the system were obtained from the following Langrangian equation [15]

$$\frac{d}{dt}\left(\frac{\partial L}{\partial \dot{q_k}}\right) - \frac{\partial L}{\partial q_k} = Q_k. \tag{1}$$

Where

$$L = T - U \tag{2}$$

T and U were obtained by summing together the kinetic and potential energies of the involved bodies, respectively. q_k are the coordinates which are coupled and dependent i.e. $(x_p, z_p, \lambda_1, \lambda_2)$.

After derivation, the equations of motion in the vertical plane are given as

$$(M_t+X_{\ddot{x}})\ddot{x}_p-M_1l_1\ddot{\lambda}_1c\lambda_1-M_2l_2\ddot{\lambda}_2c\lambda_2+M_1l_1\dot{\lambda}_1{}^2s\lambda_1 +M_2l_2\dot{\lambda}_2{}^2s\lambda_2 = \tau_{x_p} - X_{\dot{x}1}(\dot{x}_p-l_1\dot{\lambda}_1c\lambda_1) - X_{\dot{x}2}(\dot{x}_p- l_2\dot{\lambda}_2c\lambda_2) - X_{\dot{x}p}\dot{x}_p. \quad (3)$$

$$(M_t+Z_{\ddot{z}})\ddot{z}_p+M_1l_1\ddot{\lambda}_1s\lambda_1+M_2l_2\ddot{\lambda}_2s\lambda_2+M_1l_1\dot{\lambda}_1{}^2c\lambda_1 +M_2l_2\dot{\lambda}_2{}^2c\lambda_2+gM_t-\rho gV_t = \tau_{z_p}-Z_{\dot{z}1}(\dot{z}_p+l_1\dot{\lambda}_1s\lambda_1) - Z_{\dot{z}2}(\dot{z}_p+l_2\dot{\lambda}_2s\lambda_2) - Z_{\dot{z}p}\dot{z}_p. \quad (4)$$

$$M_1l_1{}^2\ddot{\lambda}_1 - M_1l_1c\lambda_1\ddot{x}_p + M_1l_1s\lambda_1\ddot{z}_p + (gM_1 - \rho gV_1)l_1s\lambda_1 = \tau_{\lambda_1} - (X_{\dot{x}1} + Z_{\dot{z}1})l_1{}^2\dot{\lambda}_1-X_{\dot{x}1}l_1\dot{x}_pc\lambda_1+Z_{\dot{z}1}l_1\dot{z}_ps\lambda_1. \quad (5)$$

$$M_2l_2{}^2\ddot{\lambda}_2 - M_2l_2c\lambda_2\ddot{x}_p + M_2l_2s\lambda_2\ddot{z}_p +(gM_2 - \rho gV_2)l_2s\lambda_2 = \tau_{\lambda_2} - (X_{\dot{x}2} + Z_{\dot{z}2})l_2{}^2\dot{\lambda}_2-X_{\dot{x}2}l_2\dot{x}_pc\lambda_2+Z_{\dot{z}2}l_2\dot{z}_ps\lambda_2. \quad (6)$$

After rearranging the terms, the equations of motion can be written in the vectorial form as

$$\boldsymbol{M}(\boldsymbol{\eta})\ddot{\boldsymbol{\eta}} + \boldsymbol{D}(\boldsymbol{\eta},\dot{\boldsymbol{\eta}})\dot{\boldsymbol{\eta}} + \boldsymbol{g}(\boldsymbol{\eta}) = \boldsymbol{\tau}. \quad (7)$$

Where

$$\boldsymbol{M}(\boldsymbol{\eta}) = \begin{bmatrix} M_t+X_{\ddot{x}} & 0 & -M_1l_1c\lambda_1 & -M_2l_2c\lambda_2 \\ 0 & M_t+Z_{\ddot{z}} & M_1l_1s\lambda_1 & M_2l_2s\lambda_1 \\ -M_1l_1c\lambda_1 & M_1l_1s\lambda_1 & M_1l_1{}^2 & 0 \\ -M_2l_2c\lambda_2 & M_2l_2s\lambda_2 & 0 & M_2l_2{}^2 \end{bmatrix}. \quad (8)$$

$$\boldsymbol{D}(\boldsymbol{\eta},\dot{\boldsymbol{\eta}}) = \begin{bmatrix} D_{t_x} & 0 & \begin{matrix} M_1l_1s\lambda_1\dot{\lambda}_1 \\ -X_{\dot{x}1}l_1c\lambda_1 \end{matrix} & \begin{matrix} M_2l_2s\lambda_2\dot{\lambda}_2 \\ -X_{\dot{x}2}l_2c\lambda_2 \end{matrix} \\ 0 & D_{t_z} & \begin{matrix} M_1l_1c\lambda_1\dot{\lambda}_1 \\ +Z_{\dot{z}1}l_1s\lambda_1 \end{matrix} & \begin{matrix} M_2l_2c\lambda_2\dot{\lambda}_2 \\ +Z_{\dot{z}2}l_2s\lambda_2 \end{matrix} \\ -X_{\dot{x}1}l_1c\lambda_1 & Z_{\dot{z}1}l_1s\lambda_1 & D_{t_1}l_1{}^2 & 0 \\ -X_{\dot{x}2}l_2c\lambda_2 & Z_{\dot{z}2}l_2s\lambda_2 & 0 & D_{t_2}l_2{}^2 \end{bmatrix}. \quad (9)$$

$$D_{t_x} = X_{\dot{x}1} + X_{\dot{x}2} + X_{\dot{x}p}. \quad (10)$$

$$D_{t_z} = Z_{\dot{z}1} + Z_{\dot{z}2} + Z_{\dot{z}p}. \quad (11)$$

$$D_{t_1} = X_{\dot{x}1} + Z_{\dot{z}1}. \quad (12)$$

$$D_{t_2} = X_{\dot{x}2} + Z_{\dot{z}2}. \quad (13)$$

$$\boldsymbol{g}(\boldsymbol{\eta}) = \begin{bmatrix} 0 \\ gM_t - B_t \\ (gM_1-B_1)l_1s\lambda_1 \\ (gM_2-B_2)l_2s\lambda_2 \end{bmatrix}. \quad (14)$$

$$\boldsymbol{\tau} = \begin{bmatrix} \tau_{x_p} \\ \tau_{z_p} \\ \tau_{\lambda_1} \\ \tau_{\lambda_2} \end{bmatrix}. \quad (15)$$

$\boldsymbol{\eta}$ is the position vector of payload and swing angles of the links, given as

$$\boldsymbol{\eta} = \begin{bmatrix} x_p \\ z_p \\ \lambda_1 \\ \lambda_2 \end{bmatrix}. \quad (16)$$

The positions of the individual HAUVs can be worked out as

$$\begin{bmatrix} x_i \\ z_i \end{bmatrix} = \begin{bmatrix} x_p - l_is\lambda_i \\ z_p - l_ic\lambda_i \end{bmatrix} \quad i = 1,2. \quad (17)$$

The states of the system are $\boldsymbol{\eta}$ and $\dot{\boldsymbol{\eta}}$. The dynamics of the system is the change of the state of the system which is written as

$$\dot{\boldsymbol{x}} = \begin{bmatrix} \dot{\boldsymbol{\eta}} \\ \ddot{\boldsymbol{\eta}} \end{bmatrix} = \begin{bmatrix} \boldsymbol{v} \\ \boldsymbol{M}(\boldsymbol{\eta})^{-1}(\boldsymbol{\tau} - \boldsymbol{D}(\boldsymbol{\eta},\dot{\boldsymbol{\eta}})\dot{\boldsymbol{\eta}} - \boldsymbol{g}(\boldsymbol{\eta})) \end{bmatrix}. \quad (18)$$

a. Propulsion Model

The propulsion model for each vehicle was developed using the principle of virtual works [16]. The generalised equation is given as

$$\tau_k = \sum_{j=1}^{4}\sum_{i=1}^{2} \frac{\partial(\boldsymbol{f}_j.\boldsymbol{\xi}_j)}{\partial q_k} \quad k = 1,2,3,4. \quad (19)$$

j represents the number of thrusters on each HAUV. i is the vehicle number and k shows the coupled and dependent coordinates of FCTS i.e. $(x_p, z_p, \lambda_1, \lambda_2)$.

$\boldsymbol{f}_j$ is the thrust force vector of each thruster. $\boldsymbol{\xi}_j$ is the position of each thruster w.r.t O, given as

$$\boldsymbol{\xi}_j = \boldsymbol{\xi}_Q + L_{Qj}\boldsymbol{e}_j. \quad (20)$$

$\boldsymbol{L}_{Qj}$ is the distance of the centre of each thruster from the centre of the vehicle. $\boldsymbol{e}_j$ is the unit vector to get the components of L_{Qj} along the payload coordinates. $\boldsymbol{\xi}_Q$ is the position of the centre of HAUV w.r.t the payload, therefore,

$$\boldsymbol{\xi}_Q = \boldsymbol{\xi}_p + l\boldsymbol{\rho}. \quad (21)$$

$\boldsymbol{\xi}_p$ is the position of the payload in EFF, l is the length of the flexible link and $\boldsymbol{\rho}$ is the unit vector along l which is given as

$$\boldsymbol{\rho} = \begin{pmatrix} -s\lambda \\ -c\lambda \end{pmatrix}. \quad (22)$$

After derivation, the thrust allocation matrices for the HAUVs are given as

$$\boldsymbol{T}_{a1} = \begin{bmatrix} 1 & 1 & 0 & 0 \\ 0 & 0 & 1 & 1 \\ -l_1c\lambda_1 & -l_1c\lambda_1 & l_1s\lambda_1 & l_1s\lambda_1 \\ 0 & 0 & 0 & 0 \end{bmatrix}. \quad (23)$$

$$\boldsymbol{T}_{a2} = \begin{bmatrix} 1 & 1 & 0 & 0 \\ 0 & 0 & 1 & 1 \\ 0 & 0 & 0 & 0 \\ -l_2c\lambda_2 & -l_2c\lambda_2 & l_2s\lambda_2 & l_2s\lambda_2 \end{bmatrix}. \quad (24)$$

$$\boldsymbol{T}_a = [\boldsymbol{T}_{a1}\ \boldsymbol{T}_{a2}]. \quad (25)$$

$$\boldsymbol{\tau} = \boldsymbol{T}_a\boldsymbol{f}. \quad (26)$$

$$\boldsymbol{f} = (f_1\ f_2\ f_3\ f_4\ f_5\ f_6\ f_7\ f_8)^t. \quad (27)$$

All the parameters for the Minerva HAUV were obtained from [14]. The moment terms were then modified by taking their effect about O. On the other hand, the hydrodynamic parameters were calculated for the payload using the Morison's formula [17] where the coefficients were obtained from [18]. The hydrostatic parameters have to be selected.

The detailed parameters are shown in Table 2.

Table 2: Model parameters

S. No.	Parameters	Values
1.	$M_1 = M_2$	520Kg
2.	M_p	To be selected
3.	$V_1 = V_2$	0.49m^3
4.	V_p	1 m^3
5.	$l_1 = l_2$	3m
6.	$I_{y1} = I_{y2}$	104Kgm^2
7.	$X_{\dot{x}_1} = X_{\dot{x}_2}$	292Kg/m
8.	$X_{\dot{x}_p}$	589 Kg/m
9.	$Z_{\dot{z}_1} = Z_{\dot{z}_2}$	635Kg/m
10.	$Z_{\dot{z}_p}$	589 Kg/m
11.	$X_{\ddot{x}_1} = X_{\ddot{x}_2}$	293Kg
12	$X_{\ddot{x}_p}$	697Kg
13	$Z_{\ddot{z}_1} = Z_{\ddot{z}_2}$	302Kg
14	$Z_{\ddot{z}_p}$	697Kg

4. SIMULATION MODEL

The simulation model was developed using the 4rth order Range-Kutta method which provides an approximate solution by numerically integrating ordinary differential equations.

The dynamics of the system is the change of states of the system which is represented at the previous time is represented as

$$\dot{\boldsymbol{x}}_n = f(\boldsymbol{x}_n). \quad (28)$$

$\boldsymbol{x}_n$ is the state vector of the system at the previous time step, given as

$$\boldsymbol{x}_n = [\boldsymbol{\eta}_n \quad \dot{\boldsymbol{\eta}}_n]^T. \quad (29)$$

$$f(\boldsymbol{x}_n) = \begin{bmatrix} \boldsymbol{v}_n \\ \boldsymbol{M}(\boldsymbol{\eta}_n)^{-1}(\boldsymbol{\tau}_n - \boldsymbol{D}(\boldsymbol{\eta}_n, \dot{\boldsymbol{\eta}}_n)\dot{\boldsymbol{\eta}}_n - \boldsymbol{g}(\boldsymbol{\eta}_n)) \end{bmatrix}. \quad (30)$$

Runge-Kutta methods are used to approximate the solutions of Ordinary Differential Equations (ODE). According to the 4rth order Runge-Kutta method, the state vector at the next time step becomes

$$\boldsymbol{x}_{n+1} = \boldsymbol{x}_n + \frac{\Delta t}{6}(\boldsymbol{k}_1 + 2\boldsymbol{k}_2 + 2\boldsymbol{k}_3 + \boldsymbol{k}_4). \quad (31)$$

Where $\boldsymbol{k}_1 = f(\boldsymbol{x}_n)$ which is the function of the state vector at the previous time. $\boldsymbol{k}_2$ is worked out from $\boldsymbol{k}_1$, $\boldsymbol{k}_3$ from $\boldsymbol{k}_2$ and $\boldsymbol{k}_4$ from $\boldsymbol{k}_3$. They are given as

$$\boldsymbol{k}_2 = f\left(\boldsymbol{x}_n + \frac{\Delta t}{2}\boldsymbol{k}_1\right). \quad (32)$$

$$\boldsymbol{k}_3 = f\left(\boldsymbol{x}_n + \frac{\Delta t}{2}\boldsymbol{k}_2\right). \quad (33)$$

$$\boldsymbol{k}_4 = f\left(\boldsymbol{x}_n + \frac{\Delta t}{2}\boldsymbol{k}_3\right). \quad (34)$$

5. INITIAL VERIFICATION

To verify the simulation code, several tests were performed.

a. Test 01

To observe the effect of restoring swing moment, the whole system was considered neutrally buoyant and buoyancy of HAUVs was ignored. No vertical thrust was required to maintain the vertical position. These conditions were assumed to verify the swing angle response by the analytical approach.

Analytical calculation: With the link length of 3m, kept initially at an angle of 40°, the restoring moment produced is 9837Nm. The same amount of moment is required in opposite to maintain the swing angles of the flexible link.

The thrusters on Minerva HAUV have a force limit of 340N. The axial thrusters are placed at an angle of 10° with the horizontal. Each axial thruster could produce a maximum thrust force of 335N. The two axial thrusters could produce a maximum thrust moment of 2009Nm for the cable length of 3m. Therefore, the angle of 40° cannot be achieved.

The swing angle which can be achieved by HAUV is analytically worked out as

$$\lambda = sin^{-1}\left(\frac{2009}{Wl}\right) = 7.5^o. \quad (35)$$

Simulation run: The simulation has also been performed under the same assumed conditions. The axial thrusters were applied at -335N and +335N on the two HAUVs, respectively. From Figure 2(a), after small initial oscillations, the swing angles of ±7.5° have been achieved for the flexible links which are exactly equal to the analytical results. From Figure 2(b), the payload does not move in surge due to the application of equal and opposite thrust forces. Whereas, HAUV01 and HAUV02 move to -0.4m and +0.4m, respectively, from the initial zero position which produces the above-mentioned swing angles. On the other hand, the payload and HAUVs are maintaining their initial vertical positions.

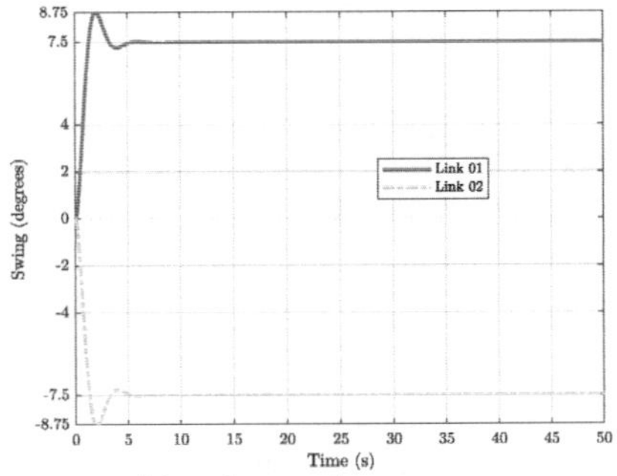

(a) swing angle response

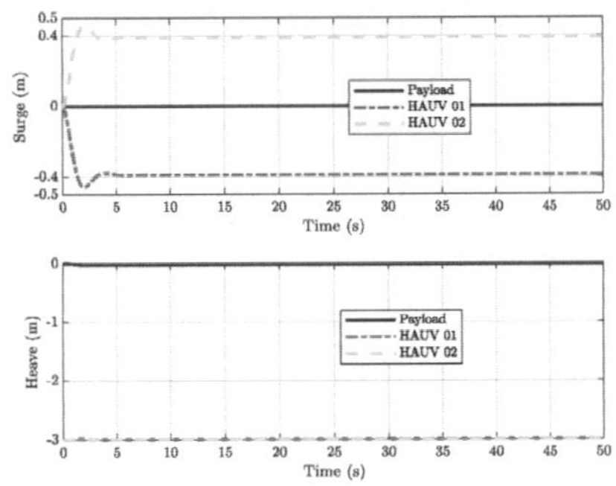

(b) displacement response

Figure 2: Test 01 – System's response

In actual, HAUVs are neutrally buoyant, therefore, there would not be any restoring moments for which the opposite thrust moments are required. This saves a lot in terms of power consumption and shows the benefit of underwater buoyancy for the flexible connection transportation system. This is even better than the flexible connection aerial transportation as the air buoyancy is very small. Therefore, autonomous aerial vehicles are kept lightweight. Otherwise, the power requirement would be higher as extra power will be wasted to maintain the swing angles of flexible links.

b. Test 02

A test was also performed keeping both the HAUVs and payload neutrally buoyant. There is no analytical solution for getting the swing angles in this case as the denominator of equation (35) becomes zero. In the simulation, the axial thrusters were applied at the operating range of -300N and +300N on the two HAUVs, respectively.

From Figure 3(a), the swing angles start increasing due to the application of axial thrusts. After 10secs, they stop at +90° and -90° for link 01 and link 02, respectively. This is practically right as the HAUVs start moving away from each other due to the application of opposite axial thrusts. As a result, the swing angles of the flexible links are increasing unresisted as there are no opposing restoring moments due to the neutral buoyancy of HAUVs. When the links reach ±90°, HAUVs get aligned with each other and with the payload. Any further opposite thrust forces would only negate their effect.

From Figure 3(b), the payload moves vertically up i.e. negative in heave from the initial position of (0m,0m). HAUV01 moves negative in surge and positive in heave from the initial position of (0m,-3m) and HAUV02 moves positive is both surge and heave from the initial position of (0m,-3m). After 10secs, the payload, HAUV01 and HAUV02 come horizontally aligned at (0m, -2m), (-3m, -2m) and (3m, -2m), respectively.

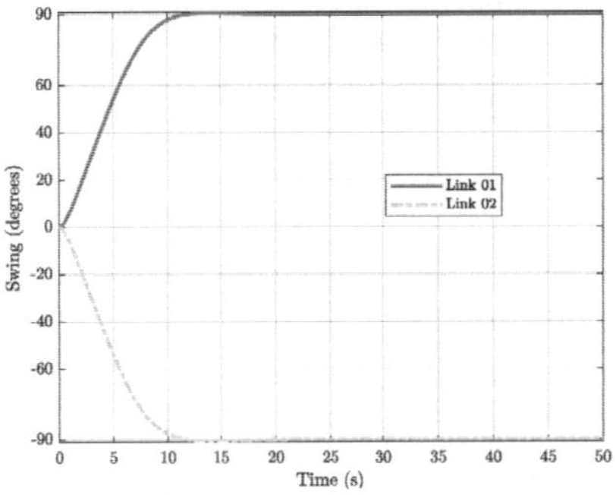

(a) swing angle response

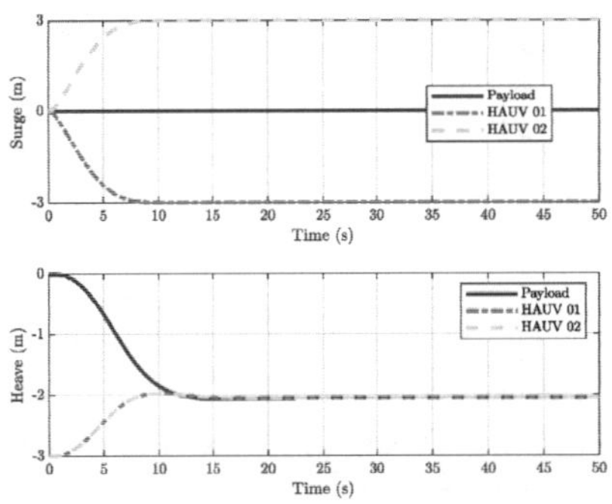

(b) displacement response

Figure 3: Test 02 – System's response

c. Test 03

Considering that the HAUVs and payload are neutrally buoyant, a test was performed in which the axial thrusters were applied at the same thrust values as applied in test 02 and vertical thrusters were applied at the operating value of 300N each on the two HAUVs.

From Figure 4(a), the swing angles of the flexible links settle at ±109° instead of ±90°. From Figure 4(b), the HAUVs move away from each other in surge and stops at ±3m, where swing angles reach their maximum values. HAUVs move above the payload in heave. Whereas, the payload only move a small distance in heave until the flexible links reach their maximum swing angles. This is because the downward motion of HAUVs do not have a significant effect on the motion of payload until they move to a point where they pull the payload and increase its rate of motion. This is not acceptable as it can cause instability of the payload as well as make it hard to get the desired swing angles of the flexible links. It is, therefore, recommended to keep the weight of payload higher than its buoyancy.

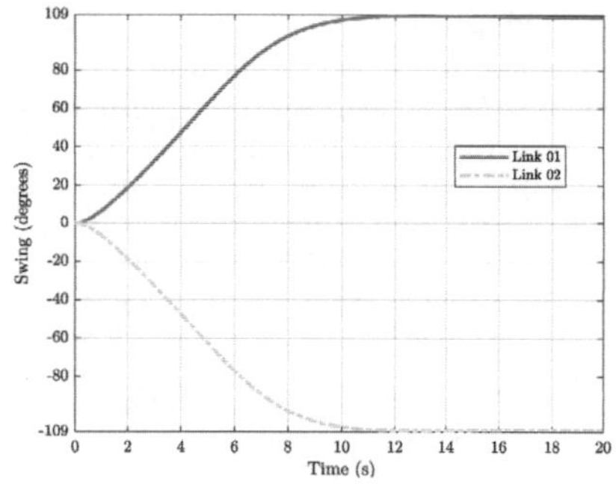

(a) swing angle response

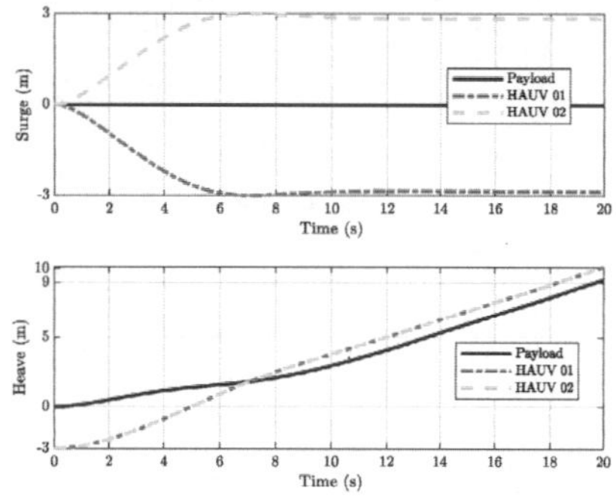

(b) displacement response

Figure 4: Test 03 – System's response

6. SELECTION OF PAYLOAD MASS

It is required to decide on the mass of payload at which appropriate swing angles of the flexible links could be achieved. Several tests were performed at different masses of the payload.

a. Test 04

The payload mass was first increased to 1050Kg which makes the weight of 10297N. The difference between weight and buoyancy becomes 245N. The axial thrusters on HAUV 01 and HAUV 02 were applied at -300N and +300N, respectively. Whereas the vertical thrusters on both the HAUVs were applied at 300N each.

Figure 5(a), the swing angles of the flexible links increases to ±100°. This is lower than the swing angles when the payload was considered neutrally buoyant in test 03. The payload starts moving vertically down in heave due to the difference in its weight and buoyancy. However, the rate of downward motion only increases when the swing angles of the links reaches their maximum values. The vertical distance travelled by the payload is increased to 12m compared to 9m in test 03.

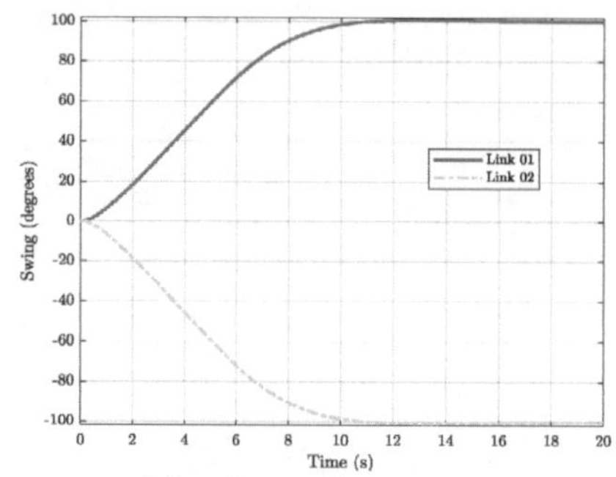

(a) swing angle response

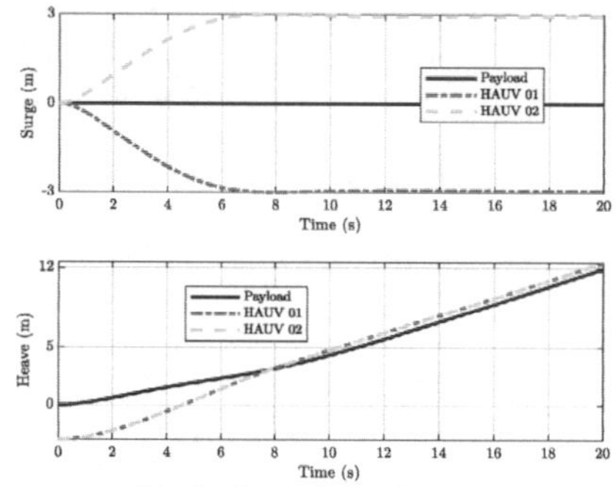

(b) displacement response

Figure 5: Test 04 – System's response

It is obvious from test 04 that a slight increase in the payload weight above buoyancy would not help in stopping the rise of the swing angles of the flexible links. This is because the rate at which the payload moves vertically down is lower than the rate at which the swing angles increase. One way is to manipulate the thrust forces in both surge and heave. However, it is difficult without the implementation of a control system. The other way is to further increase the mass of the payload.

b. Test 05

The mass of the payload was increased to 1300Kg. This makes the difference between its weight and buoyancy to 2670N. The thrusters were applied at the same thrust values as in test 04.

From Figure 6(a), the swing angles are increased to ±40° which are quite desirable. The payload moves under its weight at an appropriate rate and HAUVs remain above the payload, as shown in Figure 6(b). The payload moves a higher distance of 38m compared to test 03 and test 04.

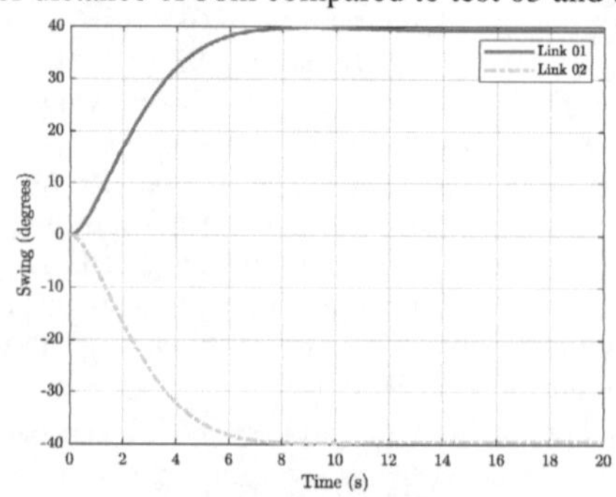

(a) swing angle response

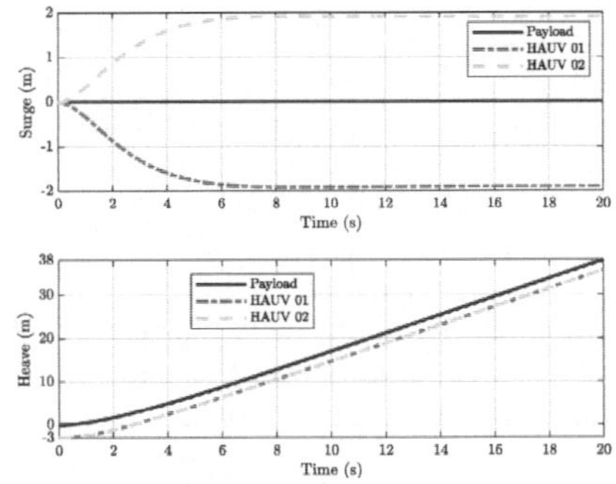

(b) displacement response

Figure 6: Test 05 – System's response

c. Test 06

The mass of 1300Kg for the payload seems appropriate when the system moves vertically down (positive heave). However, this mass is also required to be tested for the motion vertically up (negative heave). Therefore, a test was performed by applying the vertical thrusters at -300N each and the axial thrusters at the same thrust values as in test 04.

From Figure 7(a), the swing angles are increased to ±72° at which HAUVs move to a distance of ±2.9m, as shown in Figure 7(b). Though the payload keeps its position below HAUVs, the total distance travelled by it is reduced to 12.6m.

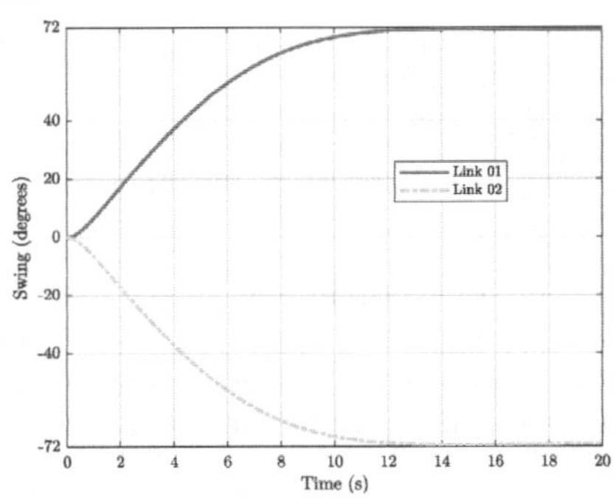

(a) swing angle response

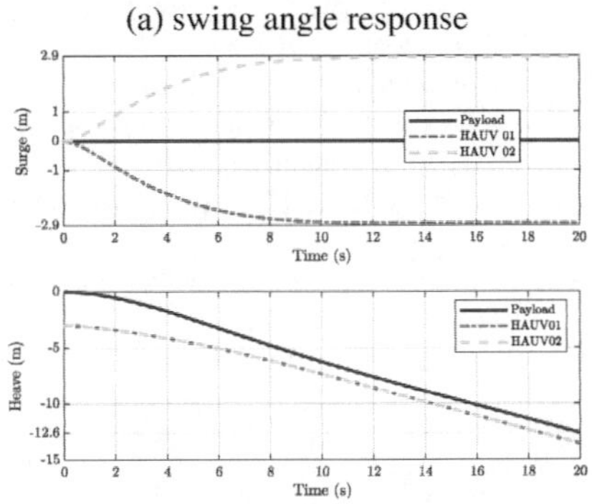

(b) displacement response

Figure 7: Test 06 – System's response

From this test, it is found that downward and upward thrust forces of the same values would produce different swing angles of the links and distance travelled by the payload. Therefore, it is difficult and hectic to decide on the thrust forces applied by the thrusters to get the desired motion response in each direction. Therefore, the design of an appropriate control system is necessary.

7. CONTROL SYSTEM DESIGN AND TRAJECTORY TRACKING

The control system was designed keeping the payload mass of 1030Kg which makes the weight of the payload slightly higher than its buoyancy. This would save a lot in power consumption. The motion response is not affected as the control system is robust enough to apply appropriate forces and vary them accordingly to maintain the swing angles while following the desired trajectory.

The control system was first designed using the PID controllers. The control system design is shown in Figure 8. The controllers are applied on the difference between $\boldsymbol{\eta}$ and $\boldsymbol{\eta_d}$. The outcome is given as $\boldsymbol{\tau_d}$ which is multiplied by the inverse of $\boldsymbol{T_a}$ to get $\boldsymbol{f_d}$. The saturation limits are applied to get $\boldsymbol{f}$ which is then multiplied by $\boldsymbol{T_a}$ to get $\boldsymbol{\tau}$. This is given as control input to the dynamic model of the system to get $\boldsymbol{\eta}$. The process continues until the desired states of the system are achieved.

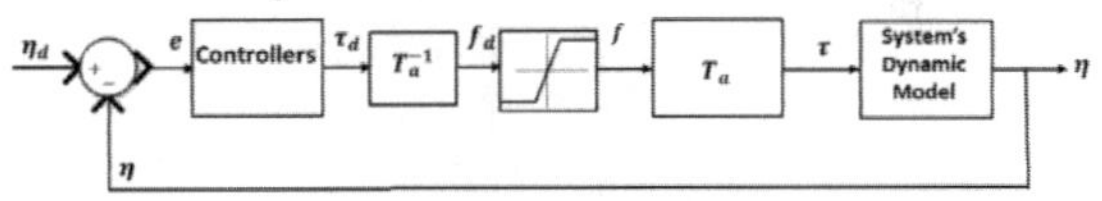

Figure 8: Control system

It was observed during simulation runs that increasing the proportional gain for surge and heave improves the payload response, but the swing angles of the flexible links are compromised, and vice versa. Therefore, two PID attempts are presented at different proportional gains for surge and heave, as shown in Table 3.

Table 3: PID gains for the system following the trajectory

		K_P	K_I	K_D
Attempt 01	Surge	100	0.1	5
	Heave	100	0.1	5
	Swing angle 01	500	10	50
	Swing angle 02	500	10	50
Attempt 02	Surge	200	0.1	5
	Heave	200	0.1	5
	Swing angle 01	500	10	50
	Swing angle 02	500	10	50

The pole placement controller was also designed to get a better comparison. To accomplish this, a state-space model of the system was developed by linearising the equations of motion at the operating point i.e. ±40° swing angles and zero position of the payload. The state-space model is written as

$$\dot{\boldsymbol{x}} = \boldsymbol{Ax} + \boldsymbol{Bu}. \quad (36)$$

$$\boldsymbol{y} = \boldsymbol{Cx}. \quad (37)$$

According to the linear state feedback control law, the control input $(\boldsymbol{u})$ is given as

$$\boldsymbol{u} = -\boldsymbol{Kx}. \quad (38)$$

Therefore,

$$\dot{\boldsymbol{x}} = (\boldsymbol{A} - \boldsymbol{BK})\boldsymbol{x}. \quad (39)$$

The pole placement control gain matrix ($\boldsymbol{K}$) is required to be worked out such that the closed-loop system has the desired eigenvalues (poles). However, it is only possible if the system is controllable.

The system was found to be controllable as the rank of the controllability matrix i.e. $[\boldsymbol{B}\ \boldsymbol{AB} \ldots . \boldsymbol{A}^7\boldsymbol{B}]$ is equal to the number of states of the system [19].

However, the system is not stable for the swing angles of the flexible links as worked out from the eigenvalues of matrix $\boldsymbol{A}$, therefore, it would not maintain the swing angles without control input.

As the system is controllable, the pole placement is possible. The poles were placed so that the system becomes stable under control input. The poles were placed at the locations as shown in Table 4.

Table 4: Poles for the pole placement controller

Sates	P
x_p	-1
z_p	-2
λ_1	-10
λ_2	-10
$\dot{x}_p$	-3
$\dot{z}_p$	-3
$\dot{\lambda}_1$	-4
$\dot{\lambda}_2$	-4

$\boldsymbol{K}$ can be worked out from the following characteristic equation

$$|\boldsymbol{PI} - (\boldsymbol{A} - \boldsymbol{BK})| = \boldsymbol{0}. \quad (40)$$

In MATLAB, 'place' command is used to work out $\boldsymbol{K}$, given as

```
K = place(A,B,P);
```
(41)

$\boldsymbol{K}$ is a matrix of the size $n \times n$, where n is equal to the number of states of the system. $\boldsymbol{K}$ is applied on the nonlinear system to get the desired motion response.

The control system design for the pole placement controller is shown in Figure 9. $\boldsymbol{K}$ is applied on $\boldsymbol{e}$ to get $\boldsymbol{f_d}$. The saturation limit is applied to get $\boldsymbol{f}$ which is multiplied by $\boldsymbol{T_a}$ to get $\boldsymbol{\tau}$. This is given to the dynamic model of the system to get $\boldsymbol{\eta}$. $\boldsymbol{\eta_d}$ is then subtracted from $\boldsymbol{\eta}$ to get $\boldsymbol{e}$. The process continues till the desired motion response is achieved.

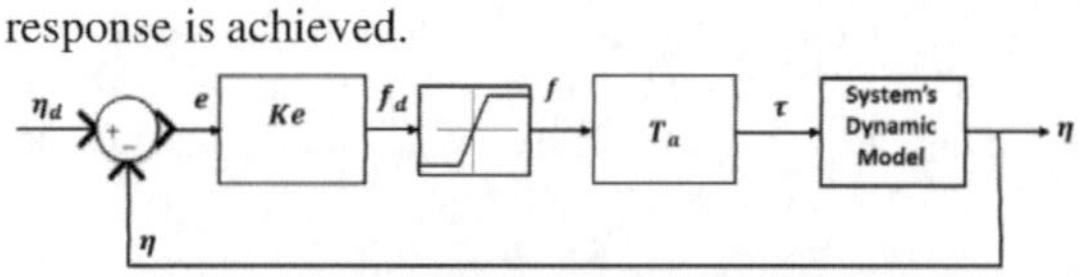

Figure 9: Control system design using pole placement controller

The control systems were implemented on the system to follow a minimum snap trajectory [6] while getting the desired swing angles of the thrusters. The waypoints of the minimum snap trajectory are shown in Table 5.

Table 5: Waypoints for the minimum snap trajectory

Waypoints	Position $x(m), z(m)$
1	(0,0)
2	(5,10)
3	(10,15)
4	(20,20)

The error of the actual trajectory tracking from the desired trajectory was calculated using the Root Mean Square Error (*RMSE*), which is given as

$$RMSE = \sqrt{\sum_{i=1}^{n} \frac{(\eta_{id} - \eta_i)^2}{n}}. \quad (42)$$

Where η_{id} is the desired state and η_i is the respective actual state of the system at step i. n is the total number of data extracted which was equal to the number of time steps in our case as the data was extracted each time step.

The power consumption by the system for each controller was worked out using the Root Mean Square (*RMS*), given as

$$RMS = \sqrt{\sum_{i=1}^{n} \frac{W_i^{\,2}}{n}}. \quad (43)$$

Where W_i is the power consumed at step i and n is the total number of steps.

From Figure 10(a), the system undershoots the trajectory in both the PID attempts. In Attempt 02, the system is more precisely following the trajectory compared to attempt 01. This is also verified by the *RMSE* values in Table 6. On the other hand, the system overshoots the trajectory when the pole placement controller is implemented. The *RMSE* values show that the error of the system with pole placement controller is higher in surge but lower in heave compared to the PID controllers.

From Figure 10(b), the swing angle response of the flexible links is closer to the desired response in attempt 01 of the PID controllers compared to attempt 02. This can also be seen in the *RMSE* values in Table 6. The swing angle of link 02 is increasing above 10° for both the PID attempts even at using high gains. On the other hand, the *RMSE* is slightly higher for link 01 but well below for link 02 when the pole placement controller was implemented. The advantage of the pole placement controller is that the *RMSE* for both the swing angles is below 10°.

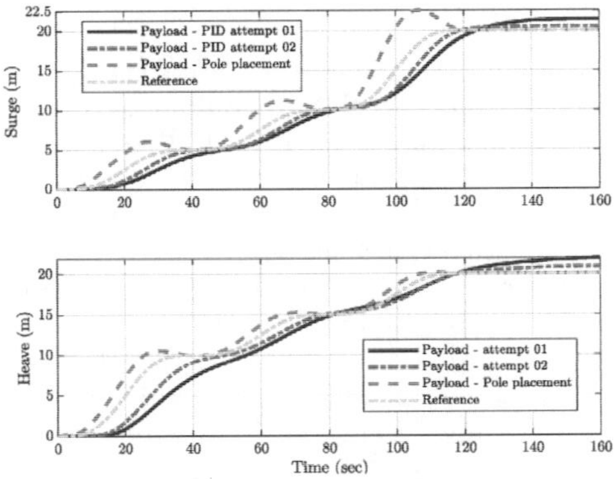

(a) displacement response

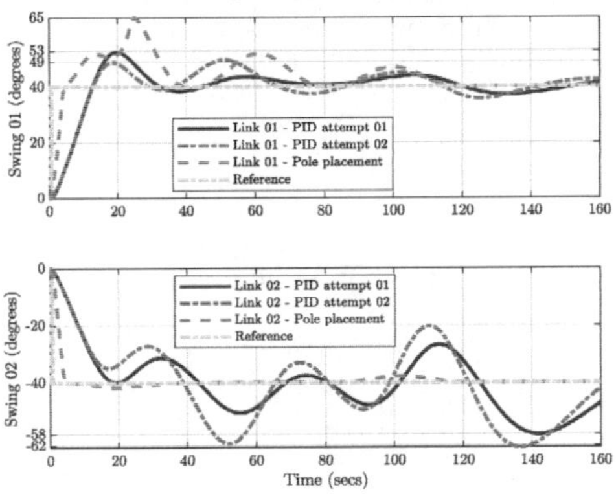

(b) swing angle response

Figure 10: System's response when following the desired trajectory

From Figure 11(a), the thrust forces which are applied by the thrusters on HAUVs are lower in PID attempt 01 compared to attempt 02. On the other hand, the thrust forces applied for the pole placement controller are significantly higher compared to the PID attempts. The maximum axial thrust forces in attempt 01 and 02 are 495N and 600N, whereas, the maximum vertical thrust forces in attempt 01 and 02 are 596N and 785N respectively. On the other hand, the axial thrust force varies between 1035N and -606N and the vertical thrust force between 1200N and -1200N for the pole placement controller.

The swing moment of the flexible links is also higher in the pole placement controller than the PID controller, as shown in Figure 11(b).

The Root Mean Square (*RMS*) values in *Table 7* show that the power consumption by the system in PID attempt 02 is significantly higher than the power consumption in attempt 01. The percentage increase is 45%. However, the power consumption further increases by 45% when the pole placement controller is applied.

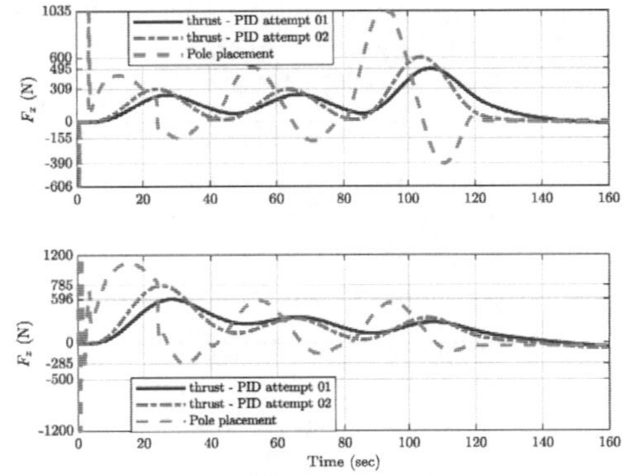

(a) force response

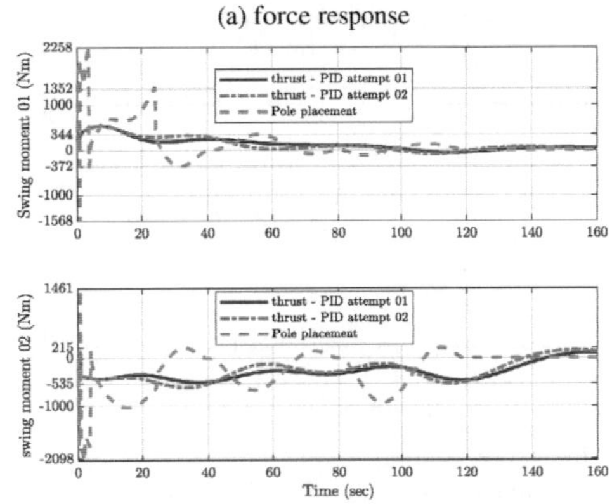

(b) swing moment response

Figure 11: thrust response of FCTS while following the desired trajectory

Table 6: RMSE of the motion response of the system and desired trajectory under different controllers

	PID Attempt 01	PID Attempt 02	Pole placement
$RMSE_x$ (m)	1.62	1.08	1.87
$RMSE_z$ (m)	2.12	1.38	0.98
$RMSE_\theta$ (°)	-	-	-
$RMSE_{\lambda 1}$ (°)	7.63	7.8	8.01
$RMSE_{\lambda 2}$ (°)	11.04	13.74	3.98

Table 7: RMS of the power consumption

	PID Attempt 01	PID Attempt 02	Pole placement
RMS_X (Watt)	61.78	161.62	208.00
RMS_Z (Watt)	70.82	94.22	184.98
RMS_M (Watt)	-	-	-
RMS_{Q1} (Watt)	8.81	2.62	51.10
RMS_{Q2} (Watt)	7.74	11.88	48.09
Total (Watt)	149.15	270.34	492.17
Percent increase (%)	14	45	45

8. CONCLUSIONS

In this paper, a nonlinear dynamic model was developed for the underwater transportation system in which two Minerva HAUVs were connected to a cubic payload via flexible links. In the initial verification tests, the advantage of buoyancy in saving the power consumption for the underwater vehicles was understood. Subsequently, it was found that the payload cannot be kept positively or neutrally buoyant as it would cause instability of the payload and would affect the motion response of the system. The selection of an appropriate payload weight was tried but the swing angles of the flexible links could not be maintained for the motion in all directions without the implementation of the control system. The control system was designed using the PID and pole placement controllers. A trade-off was found between power consumption and motion precision. The PID controllers were found to consume less power but motion precision is compromised and vice versa for the optimal pole placement controller. Amongst the two attempts of PID controllers, though attempt 01 had a smaller proportional gain in surge and heave, the power consumption was low with getting better control of swing angles of the flexible links. Only the error in the displacement response was higher but it was not significant. Therefore, attempt 01 amongst the PID controllers is recommended. This study provides a detailed analysis of the flexible connection transportation system with and without control system implementation. The selection of the type of controller in a real situation is completely the designer discretion and the user requirement. In future, the disturbances and noise will be added for more practical analysis.

9. REFERENCES

[1] M. N. Ahmadabadi and E. Nakano, "A 'constrain and move' approach to distributed object manipulation," *IEEE Trans. Robot. Autom.*, vol. 17, no. 2, pp. 157–172, 2001.

[2] A. Zaerpoora and M. Nili, "Distributed object transportation on a desired path based on Constrain and Move strategy," *Rob. Auton. Syst.*, vol. 50, pp. 115–128, 2005.

[3] H. Lee, H. Kim, and H. J. Kim, "Planning and Control for Collision-Free Cooperative Aerial Transportation," *IEEE Trans. Autom. Sci. Eng.*, vol. 15, no. 1, pp. 189–201, 2018.

[4] D. Mellinger, "Trajectory Generation and Control for Quadrotors," Mechanical Engineering and Applied Mechanics PhD Thesis, University of Pennsylvania, 2012.

[5] F. U. Rehman, G. Thomas, and E. Anderlini, "Development of a Simulation Platform for Underwater Transportation using two Hovering Autonomous Underwater Vehicles (HAUVs)," in *Proceedings of the 6th International Conference of Control, Dynamic Systems, and Robotics (CDSR'19)*, 2019, pp. 1–8.

[6] F. U. Rehman, G. Thomas, and E. Anderlini, "Centralized Control System Design for Underwater Transportation using two Hovering Autonomous Underwater Vehicles (HAUVs)," in *IFAC-PapersOnLine*, 2019, vol. 52, no. 11, pp. 13–18.

[7] M. Bernard and K. Kondak, "Generic Slung Load Transportation System Using Small Size Helicopters," *2009 IEEE Int. Conf. Robot. Autom.*, pp. 3258–3264, 2009.

[8] Z. Li, J. F. Horn, and J. W. Langelaan, "Coordinated Transport of a Slung Load by a Team of Autonomous Rotorcraft," no. January, pp. 1–23, 2014.

[9] N. Michael, J. Fink, and V. Kumar, "Cooperative manipulation and transportation with aerial robots," *Auton. Robots*, pp. 1–14, 2010.

[10] J. Ryu and S. K. Agrawal, "Dynamics and Control of a Helicopter Carrying a Payload Using a Cable-Suspended Robot," no. October 2018, 2006.

[11] F. A. Goodarzi, D. Lee, and T. Lee, "Geometric Stabilization of a Quadrotor UAV with a Payload Connected by Flexible Cable."

[12] C. Christo, "Design of Dynamic Positioning System for ROV Minerva," 2010.

[13] E. B. Holven, "Control System for ROV Minerva 2," Marine Technology Master Thesis, Norwegian University of Science and Technology, 2018.

[14] S. Marie Mo, "Development of a Simulation Platform for ROV systems," Marine Technology Master Thesis, Norwegian University of Science and Technology, 2015.

[15] M. Guo, Y. Su, and D. Gu, "Mixed H2/H∞ Tracking Control with Constraints for Single Quadcopter Carrying a Cable-suspended Payload," no. 2014, 2015.

[16] Y. Alothman, "Using Constrained Model Predictive Control to Control Two Quadrotors Transporting a Cable-Suspended Payload," *2018 13th World Congr. Intell. Control Autom.*, pp. 228–233.

[17] J. R. Morison, M. P. O'Brien, J. W. Johnson, and S. A. Schaf, "The force exerted by surface waves on piles," *Pet. Technol.*, vol. 2, no. 5, 1950.

[18] DNV, "Modelling and Analysis of Marine Operations," *tech. rep. DNV-RP-H103*, pp. 1–91, 2010.

[19] M. Egerstedt, "the control of mobile robots," 2018. [Online]. Available: https://www.coursera.org/learn/mobile-robot.

10. AUTHORS BIOGRAPHY

Faheem Ur Rehman is currently undertaking PhD studies at UCL Mechanical Engineering department. His field of research is autonomous underwater transportation using multiple autonomous vehicles. He is a Member of the Royal Institute of Naval Architects (RINA) and a student member of the Society of Naval Architects and Marine Engineers (SNAME).

Dr Enrico Anderlini is a senior research associate at UCL Mechanical Engineering. He teaches ocean engineering at undergraduate and postgraduate level and provides support to the ship and submarine design exercises. During his doctorate, Enrico developed control strategies for wave energy converters using reinforcement learning. He is an Associate Member of the Royal Institute of Naval Architects (RINA).

Professor Giles Thomas is BMT Chair of Maritime Engineering in University College London's Department of Mechanical Engineering. A naval architect, his research focusses on the performance of ships, boats and offshore structures. Specific interests include fluid-structure interaction, hydrodynamics, full-scale measurements, model testing and design. He is a Fellow of the Royal Institution of Naval Architects (RINA) and a Chartered Engineer (CEng – Engineering Council of Great Britain).

Ahmed Ali is a senior lecturer at the Pakistan Group of Colleges, Pakistan. He is a Mathematician and a Statistician who has a wide experience in problem-solving techniques. His research focuses on the development of complex dynamic models using the Langrangian method.

REVIEW OF TECHNOLOGIES FOR REAL-TIME SHIPBOARD ICE SEVERITY SENSING

M. Garvin, National Research Council, Canada

SUMMARY

This paper contains a review of technologies employed in the literature to provide data on ice conditions that a ship may encounter in seconds to minutes. Masters use a variety of information to assess the conditions facing their ship. This information is used assess ice severity and to decide how the ship will navigate the ice confronting it while avoiding damage to the ship, besetting, etc. Typically, this assessment is made by the officer in charge with significant levels of experience needed to assess and navigate severe ice conditions. A challenge in allowing autonomy or decision-support for navigation of ice-covered waters is providing ice severity data that feeds decision-making algorithms. Since no sensor technology can provide a complete measurement, an ice sensor suite is envisioned as a set of sensors that measure the ice surrounding the ship, building a local map of ice severity, and measure the interaction between the ship and the ice, providing calibration.

1. INTRODUCTION

The work described herein was undertaken as part of a project to specify and evaluate sensor suites for use aboard vessels operating in ice to detect ice severity in real-time. Such a sensor suite is envisioned as an integral part of a decision support system and eventual autonomous ship navigation in ice since assessing ice conditions and integrating that information into routing and navigational decisions is vital to safe and efficient operation. In order to fulfil this objective, the sensor suite must be able to provide both tactical-level information (to allow routing decisions on the scale of hours to days) and operational-level information (to allow navigational decisions such as helm inputs and propulsion power on the scale of seconds to minutes). This paper will focus on sensor technologies that may provide data that is useful at an operational level. To do this, the sensor suite must provide real-time measurement or estimation of the ice severity confronting a ship.

The purpose of this paper is to provide a summary of techniques used in scientific studies and discussed in the scientific literature that involve measuring ice properties at an operational level. Taken together, these techniques are expected to have potential to provide a comprehensive evaluation of ice severity confronting ships navigating in ice-covered waters. With this information, a system of sensors may be proposed that could be deployed aboard a ship to provide real-time estimation of ice severity. This type of system would inform decision-support and eventual autonomous navigation in ice-covered waters. A companion paper is being written that investigates tactical-level technologies.

2. PROBLEM DEFINITION

Navigating in ice-covered waters presents unique complications for ship navigation. In addition to the risks of structural damage and possible loss of the ship, many ships become beset by ice and require rescue. In some cases, besetting can pose a risk to crew and cargo even if there is no structural damage to the ship, if the duration of the besetting exceeds the supplies aboard the ship. Current practices rely on the training and experience of the ships' masters and an array of forecasting tools to avoid conditions that may result in damage to the ship or besetting.

In order to estimate the severity of ice confronting a vessel, measurement of the following ice properties is desired: *Thickness, Strength, Concentration, Composition (snow, rubble, brash, ridging, refrozen rubble, 1st year, multi-year, etc.), Piece size (dimensions and mass),* and *Pressure*. In addition, the following vessel-specific measurements can provide insight into the severity of the ice confronting the vessel: *Impact load* and *Vessel propulsion power*.

There is no single sensor technology that can acquire all of the required data. The eventual system is envisioned as a suite of sensors that may be customized to specific applications, with each sensor fulfilling a subset of needs.

3. METHODOLOGY

Searches for research literature on real-time ice sensors and/or sensing techniques were undertaken in the following databases: *Scopus, Arctic & Antarctic Regions, and the National Technical Information Service*. These searches formed the starting point, returning records which were categorized and reviewed. In addition to the database searches, other publicly-available sources were considered such as publications by ice forecasting organizations.

4. TRADITIONAL ICE MEASUREMENTS

4.1 THICKNESS MEASUREMENTS

The first category of thickness measurements typically involves stopping the ship and placing people and/or equipment on the ice to drill through the ice. Common methods include drilling a series of holes along the intended route for the ship to measure centreline thickness of the track. Another common technique involves sensors such as ground-penetrating radar dragged on sleds along specific patches of unbroken ice (ie, intended routes) to profile continuous ice thickness along a single path [1].

Over the Side Video (OTSV) is a technique that involves nadir-oriented video cameras located at the bow shoulders of an icebreaking vessel. This location provides the cameras with a view of the edge of broken ice pieces as they surface edge-on alongside the ship. This technique has been used fairly extensively and, recently, has been greatly aided by the proliferation of cheap, readily available GPS-enabled video cameras. In concept, it is similar to techniques used to visually assess ice thickness but adds the benefit of providing an ongoing record. In studies with comprehensive complimentary borehole thickness measurements, OTSV showed good correlation of ice thickness with the borehole measurements while having less success with snow thickness measurement. OTSV has been more successful in first-year ice and, notionally, is expected to be more successful when ice thickness is small compared to broken piece size. It has been used successfully in pack ice and in level ice. [2]

4.2 STRENGTH MEASUREMENTS

While not without its imperfections, the in-situ cantilever test is a widely used and reliable method of measuring ice flexural strength both in the field and in a laboratory environment. In the field, these tests present a significant logistical challenge as the beam sections must be sawn out of pieces of level ice and significant equipment must be placed on the ice to apply load and measure deflection. [3]

Ice strength can be estimated from core samples by applying empirical relationships between temperature, salinity, density, and flexural strength. Immediately following the retrieval of a core, the temperature in the center of the core can be measured at intervals along its length, giving a good indication of the temperature profile through the thickness of the ice. Density is measured by measuring the force required to submerge a piece of ice in fresh water. Puck-shaped pieces are normally cut from locations along the length of the core to provide measurements of density through the thickness of the ice. Lastly, samples from the core are melted and their salinity checked. Measuring salinity of samples from a range of locations within the core allows the assessment of salinity through the thickness of the ice. Brine volume is calculated from Temperature, Density, and Salinity, using the relationship in Cox & Weeks [4]. The flexural strength, in MPa, is then calculated using the relationship in Timco & O'Brien [5].

4.3 CONCENTRATION & PIECE SIZE MEASUREMENTS

Operational shipboard concentration and piece size measurements are typically made visually from the bridge of the vessel. Where possible, this information is supplemented by aerial photography and radar data, either from independent aircraft flyovers or from balloons or drones accompanying the ship. Traditionally, these images and data have been interpreted by ice specialists or crew members aboard the ship. [2]

4.4 PRESSURE MEASUREMENTS

Pressure fields within high-concentration pack ice derive from environmental forces such as wind and current shear and thermal strain. Floes contact their neighbors and form a "lattice" that allows pressure to build. If this pressure exceeds the ability of the weakest ice floes to withstand it, the pack ice can fail, resulting in ridging and/or rafting of floes until a sufficiently strong lattice is reestablished. These areas of rafting or ridging pose a threat to shipping as ridges are a major cause of besetting but they are not the only hazard from ice pressure. The forces within the pressured ice lattice can exert extreme forces on ships, leading to besetting and structural damage. A ship passing through an area of pressured ice may sufficiently disturb the lattice to initiate rafting, ridging, or ice pile-ups against the hull. [6]

A widely-used operational assessment of relative pack ice pressure is to observe the lead of open water left in the wake of the ship. If the lead remains open, ice pressure is low but if the lead closes tightly behind the ship, ice pressure is high. Experimental evaluations of ice pressure involve mounting stress meters within large ice floes and monitoring them throughout an ice season to detect & measure pressure events. These experiments are difficult and few have been conducted. [7]

4.5 ICE AGE MEASUREMENTS

Differentiation between first-, second-, and multi-year ice is a critical concern due to the vast difference in thickness and strength between these ice types. Because multi- year ice has the potential to be so much thicker and stronger than first-year ice, it is implicated in the vast majority (75%) of incidents involving ship damage in the Canadian Arctic, with severe incidents ("large hole" or "ship sank") being nearly entirely attributed to multi-year ice [8]. In particular, smaller pieces of multi-year ice within larger areas of level or pack ice can be difficult to distinguish and present a hidden threat. Reliable measurement of ice age is based on core sampling. Cores allow for through-thickness profiling of salinity and microstructure which, together form the basis to determine the age of the ice. At a granular level, first-year ice is composed of slender needle-like ice crystals with interspersed brine and air pockets and sometimes solid salt crystals. The entrained brine and salt are a result of the formation process of the ice; as the ice crystals grow, they cannot eliminate all of the salt, trapping pockets of brine and salt crystals between the crystals. The concentration of salt within the brine pockets is sufficient to prevent further freezing. As temperatures increase, these brine pockets grow, eventually joining and, eventually, allowing brine drainage. If the remaining ice survives until the next winter, new ice will form on the underside as temperatures fall, leading second-year ice to exhibit a structure with low-salinity ice formed in the previous season with higher-salinity new ice below. The old ice is characterized by little if any brine or salt inclusion. Multi-year ice is classified by having several of these layers made up of ice that survives the summer with newer ice below. Brine drainage is progressive; older ice has the least brine and salt inclusion. The grain boundaries are also more rounded/smoothed in older ice. This leads to distinctions that can be made based on the colour of the ice with first-year ice being predominantly white and multi-year ice taking on a turquoise blue colour. Inferences about ice age can also be made by the surface appearance and roughness of floes. As the top surface of the ice is subjected to summer sunlight, melt pools form and, in many cases drain through the ice. This leads to multi-year ice exhibiting an increasing a "hill and dale" appearance with progressively interconnected drainage ponds and channels. The surface roughness of multi-year ice can be distinguished from ridging in first-year ice since the surface roughness is more rounded in multi-year floes. Multi-year floes can often be identified within a mixed ice field by their increased freeboard compared to surrounding younger ice. [9, 2]

5. RELEVANT TECHNOLOGIES

5.1 VISUAL TECHNIQUES

Visual techniques have been used to assess ice type, concentration, and piece size from images of the ice surrounding a ship as well as thickness using advances on Over-the-Side Video (OTSV) techniques. One drawback that affects all techniques using visual data is their reliance on lighting to provide clear images. Low light or poor visibility due to fog, etc. degrades thee techniques, particularly for images of ice further from the vessel, beyond the range of ship-based lighting. To date, use of OTSV has involved interpretation of the video by researchers although some work has been undertaken to automate this process using machine vision techniques [10]. This work is promising, particularly in thick ice although further work is needed. Improving the image analysis process is expected to have potential to not only speed up the analysis but also yield improved insights into the composition of the ice by examining colour variations in the ice cross sections. [11, 1]

There has been significant recent interest in analysis of image data of ice surrounding a ship to provide estimates of ice concentration and floe size based on its surface characteristic and distinguish between level ice, ice floes, brash ice, bergy bits, and icebergs. It is also promising that some of this work has been completed using open-source image classification software. [12, 13] The main problem with obtaining detailed ice measurements from image-based data is that ice presents very low colour contrast and often limited features (simply put, image processing must differentiate between varying shades of white). Because of the relatively high colour contrast between ice and open water, ascertaining the presence of ice by differentiating between ice and open water is the best-developed use of image data to date.

Stereo cameras have been used to measure the surface topography and other surface features of sea ice. Building on image processing developments in other disciplines, multi-layered image processing techniques have been developed that show promise in processing ice image data in ways that allow surface topography to be measured in low-contrast images [14]. Recent work [15] has shown that automated stereo camera systems using machine learning and convoluted neural networks for image processing can provide promising measurements as a supplement to trained observers in detection of ice presence, ridges, melt pools, and more nuanced defined topographical features but cannot yet provide real-time ice information.

5.2 LIDAR

Aircraft-borne LIght Detection And Ranging (LIDAR) has been used in several studies to measure the surface roughness of ice and has been attempted as a method to estimate ice thickness. The applications of LIDAR typically involve measurements of surface topography which are useful for detecting features such as isolated floes & ridge sails with accuracies of 5-25cm. Rivas et. al. [16] used a LIDAR-based Airborne Topographical Mapper borne on an aircraft to distinguish between first-year ice and multi-year ice based on the differences in topography discussed above. While there will be some challenges associated with the angle of incidence, it is expected that many techniques used for airborne LIDAR would be applicable to high-mounted ship-based LIDAR.

Aircraft-borne LIght Detection And Ranging (LIDAR) and Airborne Topographical Mapping (ATM) has been used in several studies to measure the surface roughness of ice and has, more recently, been attempted as a method to estimate sea ice thickness. Estimates of freeboard can be made using LIDAR if areas of open water are present to provide a sea level reference. This information, in turn is often used to provide an estimate of thickness subject to the large uncertainty driven by assumptions of ice density, etc. discussed above [17]. NASA's Operation IceBridge involved an extensive set of aerial missions with a suite of sensors including an ATM (to measure freeboard), snow radar (to measure snow depth), and photogrammetry. This dataset allows for the differentiation between snow and ice cover since the snow depth is directly measured, reducing the uncertainty of the resulting ice freeboard estimates to a vertical accuracy of 6.6cm and a vertical precision of 3cm [18]. Still, uncertainty in sea ice thickness (once sources such as ice density were included) ranged from 58-78cm for estimated ice thicknesses of 1.77-3.82m [19].

5.3 GROUND-PENETRATING RADAR & EM SENSORS

Ground-Penetrating Radar (GPR) and Electromagnetic Induction (EM) sensors are well-established technologies for measuring thickness of ice, including sea ice. Most frequently in studies of sea ice, the GPR sensor is moved across the surface of the area of interest but there are well established methods of deploying GPR and EM sensors underneath a helicopter. While the details of these methods vary, they commonly involve deploying the sensor at between 3-20m above the surface of the ice [20, 21, 22]. These results are of particular interest since the distance above the ice surface is achievable for ships with bow-mounted sensors.

EM in particular is a mature technology that has been in use since the 1980s with ongoing improvements in data collection and processing. EM measurements rely on the contrast in conductivity between ice and the underlying sea water; the transmitted EM field (kHz to 100s of kHz) penetrates the low-conductivity sea ice but generates eddy currents in the higher-conductivity sea water which are measured by the sensor, providing a reliable measurement of the distance of the ice/water interface from the sensor. Combined with a measurement of the ice top surface provided by, for instance, a laser altimeter, this technique provides good accuracy. Haas et. al. [23] showed thickness measurement accuracy of 0.1 m (dependent on sensor height) with an EM sensor flown 10-20m above the ice at 148-167km/h. Since EM measurement relies on a contrast in conductivity, it has been demonstrated over salt water and brackish water ice.

Galley et. al. [24] used 250 MHz &1 GHz GPR to measure snow and ice thickness with good resolution (within 8.3% of in-situ measurements for snow and 5.6% for sea ice). In addition to providing basic thickness data, GPR has been used to measure the thickness of different layers within the ice, typically discerning between snow layers and ice layers and has been shown to work for ice over water of low salinity [20]. Radar signals (100s of MHz to GHz) reflect off the interfaces of contrasting dielectric constant (such as the snow-ice interface and the ice-water interface). In general, better results are achieved in level ice with signal quality degrading in formations such as ridges as signals are scattered by angled ice interfaces.

GPR has been shown to be capable of providing an estimate of the salinity of sea ice by estimating the dielectric properties of snow, fresh water ice, and sea ice. Liu et. al. [25] achieved good dielectric properties measurements using broadband GPR (250MHz to 1.5GHz) and a single pair of transmitting/receiving antennas achieving good correlation with in-situ measurements. In turn, the dielectric properties may be used as a measure of brine volume in sea ice according to the relationship in Backstrom & Eicken [26], allowing an estimate of ice strength to be made according to Timco & O'Brien [5].

5.4 ICE RADAR

Recent publishing on ice radar for discrimination between ice types is scarce but earlier work showed an ability to differentiate between first year ice, multi-year ice, and icebergs based on Bayesian classification of data collected from sideways-looking radar [27]. Commercially-available ice radars now offer ridge, berg, and lead detection and can differentiate between icebergs/bergy bits and low-freeboard multi-year ice embedded within first-year pack ice [28]. More commonly, radar is used to measure presence and movement of ice. Shore- or ship-mounting radar gives a low incidence angle meaning that it is most sensitive to areas of roughness or discontinuity (eg. ridges, rubble, and floe edges) often failing to recognize areas of level, uniform ice. [29]

5.5 HULL LOADS

Several researchers have directly measured impact loads by retrofitting various styles of load sensing device to the outside of the ship's hull. These systems are generally welded directly to the hull structure and extensive effort is required to ensure they are not damaged during testing – generally they are not suitable for prolonged operational use. Primarily, these types of sensor are useful for measuring pressure over a specific region of the hull, particularly due to impacts. Generally, they are less well suited to operational load measurement over a prolonged period of time. [30, 31]

Many studies have investigated the use of strain gauges bonded to the interior of the hull structure to monitor loading. FEA techniques help to define points of interest in the hull structure and are useful in interpreting the results [32, 33]. This type of measurement is quite difficult to obtain since strain gauges must be applied to the difficult-to-access interior surfaces of the hull structure. Accessing the required areas to install or service such systems is often prohibitive and may be most easily achieved if the system is installed while the ship is being built. That said, it is possible to retrofit this type of sensor to many parts of the hull and in-situ calibrations are possible to increase confidence in the results. Similar to externally-mounted pressure panels, this technique is most suited to measuring local load in specific areas of the hull. For best results, detailed finite element analysis of the hull structure is necessary to identify optimal locations for strain gauge placement. Deployed on areas of specific interest, this type of sensor can form part of a decision-support system that allows masters to assess hull loading in near real-time [34, 33, 35].

5.6 HULL ACCELERATIONS & VIBRATION

Global hull impact loads have been measured by accelerometer-based systems which use acceleration measurements and calculated total mass (mass of ship + added mass) to estimate load as the ship decelerates when colliding with an object [36]. This technique has been used on several studies to measure ramming loads generated when an icebreaker is breaking through ridges or colliding with ridges, bergy bits and even smaller impacts such as large late-season pack ice floes [37, 11].

The concept of using hull vibration as an indicator of ice resistance is based on fundamental icebreaking theory but this method has been applied in more recent studies in near real-time to allow a ranking of the severity of ice conditions [38]. The goal of this method is to allow estimation of ice resistance during continuous ice-breaking (eg. level ice, brash ice channels) rather than looking at the loads generated by discrete impacts. This method has received significant attention in recent years with a focus on sophisticated data-processing methods that are required to interpret the data. Specifically, good results have been achieved whereby the measurement of accelerations in the bow has been able to assess encountered ice conditions and distinguish between different levels of ice coverage [39]. This assessment uses two general methods: advanced statistical tools statistical tools or machine learning algorithms. While datasets including both direct ice severity measurement and acceleration data are required for both methods to develop or calibrate ship-specific relationships, Heyn et. al. [39] indicate that statistical methods appear to be more robust and provide more reliable results. Given the recent acceleration of interest in this topic, it is reasonable to expect that techniques may be possible that can provide more nuanced assessments of ice severity [35].

5.7 AERIAL VEHICLES

Many ice breakers use helicopters extensively to observe ice conditions and assist with determining optimal routes. This type of capability, however, is beyond the scope of most vessels operating in ice. Balloons are a well-proven technology for obtaining aerial information about ice conditions surrounding a ship and ship-based aerial drones are increasing in popularity. While there are undoubtedly challenges with operating drones including regulatory restrictions, limited endurance, and limited reliability in harsh weather, drones present a low-cost option for a flexible platform to carry many of the sensor types discussed above. Several companies offer drones capable of carrying sensors such as GPR as a product or drone-based GPR surveys as a service. Drones carrying visual cameras have proliferated widely at all levels of cost and quality.

5.8 UNDERWATER VEHICLES

Remotely-Operated underwater Vehicles (ROVs) and Autonomous Underwater Vehicles (AUVs) are widely used in underwater survey, inspection, and experimental work. ROVs specifically designed to provide under-ice data have been successfully deployed up to 20 km beyond the ice edge, providing a diversity of data [40]. Sensors such as Acoustic Doppler Current Profilers (ADCP) and Ice Profiling Sonars (IPS) can be mounted to AUVs in a manner similar to how significant ice thickness data has been collected from naval submarines operating under Arctic Sea Ice [41]. A discussion of the use of these sensors is provided in the companion paper. This type of technology is especially suitable for selective use in areas where high ridging or rubbling is expected since these phenomena make thickness measurements via other techniques more difficult. The vehicle could be deployed to survey the underside of the ice along the planned path of the ship.

6. PROPOSED SENSOR SUITE

The proposed sensor suite is a critical first step in the process of providing accurate and actionable information to ships' masters through a decision support system or to an autonomous ship. This is envisioned as a process similar to that shown in Figure 1, building on that proposed by Bekker et. al. [42]. In this process, data from multiple sensors is analysed to determine a prediction of conditions in the immediate vicinity of the ship that the ship may encounter on the order of seconds to minutes (operational). This data is combined with regional data on ice conditions that the ship could encounter on the order of hours to days (strategic) and a model of the ship's performance in ice. At a basic level, this ship performance model could be as simple as an AIRSS Vessel Class or POLARIS RIV but could be more sophisticated, including for example a model of the ship's achievable speed based on ice severity. Depending on the sophistication of the ship performance model, the task of identifying and optimizing paths could range from determining "GO/NO-GO" or an assessment of risk level using AIRSS Ice Number or POLARIS RIO to a path that is optimized to balance risk level against transit speed, etc.

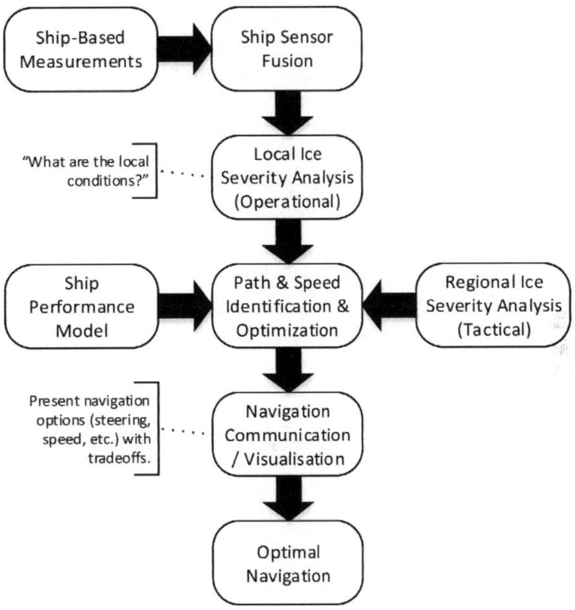

Figure 1. Process for converting data to actionable information.

The sensor suite can be viewed as two sets of sensors. The first provides estimates of ice conditions prior to interaction with the ship and the second measures the interactions themselves. While sensors in the second category are not directly applicable to planning operational navigation, they have great potential to validate any predictions of made using other sensors. This validation strategy allows refinement of ship-specific ice severity estimates by using measurements of the ship's response to ship-ice interactions (delivered power and impact load, for example) to refine a predictive model of the ship-specific local ice severity. Further to providing calibrations for onboard sensors, all shipboard sensors can be used as verification for information in ice forecasts.

With reference to the list of required measurements in section 2.1, the following technologies provide promising solutions to acquire the required data:

Concentration & Piece Size – Concentration and piece size can be measured using visual-spectrum cameras mounted high on the vessel and providing a 360 degree view of ice conditions around the ship. LIDAR and ice radar can perform similar functions and provide redundancy of this measurement. The benefit of ice radar is that it will function in restricted visibility with 360 degree cameras filling-in any blind spots closer to the ship.

Thickness – A reliable method of measuring thickness in advance of the ship is by using a GPR or similar sensor mounted on a drone. A GPR unit mounted over the bow of the ship bridges the gap between predictive and confirmation measurements. It is likely that measurements from a sensor thus situated would not provide sufficient advance warning to adjust course or speed but the data would provide ongoing confirmation of ice thickness and height above water to other sensors. As the ship contacts ice, confirmation of thickness estimates could be provided by GPR or OTSV. The ability to provide ongoing calibration using bow-mounted GPR gives rise to the possibility that freeboard measurements from a sensor such as LIDAR could prove sufficiently accurate thickness estimates for operational use.

Composition (snow, rubble, brash, ridging, refrozen rubble, 1st year, multi-year, etc.) – Features that result in changes in topography can be identified by LIDAR. Visual techniques provide the definitive measure of ice type but automated image-classification techniques require refinement. Ice radar can provide significant information, identifying ridges and multi-year ice. To provide confirmation of forward-looking measurements, a bow-mounted GPR or EM sensor can provide measurements of snow, first-year and multi-year ice.

Strength – GPR provides the most promising method of measuring ice strength in real-time. As with the thickness measurements above, GPR-based strength measurements could be made from a combination of drone-based sensors and sensors mounted over the bow. Absent forward-looking GPR data, generalizations of strength can be made based on categorization of the ice composition.

Pressure – Pressure can be measured by cameras mounted to the transom of the vessel specifically to assess how quickly the wake closes behind the ship.

Impact & icebreaking load – strain gauges on hull shell provide the most detailed information with the ability to specifically assess the likelihood of damage and overall hull structural health. Through judicious placement of these sensors, it is not only possible to measure loads in the primary icebreaking areas but also secondary loads such as athwartships pressure. The drawback of these sensors is that they measure impact loads as they are happening; too late to avoid the impact. Since extensive application of hull strain gauges is often prohibitively complex, impact load estimates from hull acceleration & vibration measurements can be used in a similar manner and likely have higher potential of widespread installation and reliability at the loss of some nuance in the data. While the technology to assess impact load from whole-body acceleration is mature, further work is needed to refine techniques for assessing continuous icebreaking force using hull vibration.

Vessel propulsion power – vessel propulsive power is relatively easily measured, especially with the proliferation of propulsion control systems that have the capability to measure and log information such as propeller shaft delivered power, thrust, torque, and rotational speed. If a direct measurement of delivered power is not available, it can be calculated from shaft torque and rotational speed. Similarly, if there is no direct measurement of thrust, it can be estimated from torque, rotational speed, and propeller characteristics. As with impact load, propulsion power is not predictive but rather measures the result of the ice the ship is currently in. Real-time measurements of delivered power could be used to validate estimates of ice severity with particular relevance to besetting.

7. CONCLUSIONS

Many technologies exist at varying levels of maturity to assess ice severity in support of operational ship navigation. The sensor suite described in Table 1 is proposed. Sensors are classified as "Predict" meaning that they provide estimates of conditions prior to interaction with the ship, with sufficient time to adjust navigation or "Response" meaning that they measure the interaction of the ice with the ship and are used to improve estimates. "*Priority 1*" sensors are proposed as the fundamental sensor package while "*Priority 2*" sensors could be added to provide more complete data.

There is considerable work to be done to make a reliable system similar to the one proposed. No one technology can provide a complete picture of ice severity and many technologies have overlapping capabilities. There is a significant sensor fusion effort required to combine the relevant measurements into an overall picture of ice severity. There are inevitable measurement uncertainties associated with even the most promising sensor types. In order to provide robust, accurate data, it is proposed that ship responses (measurement of broken ice thickness, propulsion power, impact load, icebreaking load, etc.) be measured to provide ongoing calibration of the primary operational sensors.

While measurement of ice presence and ice type/age (and, hence, expected strength) is achievable with relatively mature Ice Radar technology, this technology currently doesn't provide a complete picture. Two areas for real-time sensor development include enhanced methods of detecting ice strength (beyond classification of ice type) and more reliable methods of measuring ice thickness (including variations in thickness such as ridges).

Use	Sensor	Data
Predict	Ice Radar	Concentration Piece Size Identification of multi-year ice
Predict	LIDAR	Concentration Piece Size Ridge Identification Freeboard Ice Age
Predict	360° Camera	Piece Size Concentration
Predict / Response	Bow-mounted GPR	Thickness Composition Strength
Predict / Response	Bow-mounted EM	Thickness
Response	360° Camera	Pressure
Response	OTSV	Thickness Composition
Response	Whole Body Acceleration	Global Impact Load
Response	Propulsion Power	Total Ice Resistance
Predict	Aerial Vehicle	Concentration Piece Size
Predict	Underwater vehicle	Ridge keel dimensions Thickness
Response	Whole-body Vibration	Continuous Icebreaking Load
Response	Hull strain	Local Impact Load Structural "Health"
Priority 1 Sensors		
Priority 2 Sensors		

Table 1. Proposed Sensor Suite.

8. REFERENCES

[1] S. J. Jones, C. Kirby, C. Meadus, W. Tucker, J. Gagnon and B. Elder, "Sea ice properties on the USCGC Healy ice trials," in Proceedings of the 16th International Conference on Port and Ocean Engineering under Arctic Conditions, Ottawa, 2001.

[2] Canadian Ice Service, "Manual of Ice (MANICE) Chapter 2: Ice Observations," Environment and Climate Change Canada, 06 2005. [Online]. Available: https://www.canada.ca/en/environment-climate-change/services/weather-manuals-documentation/manice-manual-of-ice/chapter-2.html. [Accessed 05 05 2020].

[3] R. M. W. Frederking and F.-U. Hausler, "The flexural behaviour of ice from in situ cantilever beam tests," in IAHR Symposium on Ice Problems, Lulea, 1978.

[4] G. F. Cox and W. F. Weeks, "Equations for Determining the Gas and Brine Volumes in Sea Ice Samples," COLD REGIONS RESEARCH AND ENGINEERING LAB , Hanover, NH, 1982.

[5] G. W. Timco and S. O. O'Brien, "Flexural strength equation for sea ice," Cold Regions Science and Technology, pp. 285-298, 1994.

[6] I. Kubat, D. Fowler and M. Sayed, "Chapter 18 - Floating Ice and Ice Pressure Challenge to Ships," in Snow and Ice-Related Hazards, Risks and Disasters, Elsevier, 2015, pp. 647-676.

[7] G. Comfort and R. Ritch, "Field measurement of pack ice stresses," Proceedings of the International Offshore Mechanics and Arctic Engineering Symposium, vol. 4, pp. 177-181, 1990.

[8] I. Kubat and G. Timco, "Vessel Damage in the Canadian Arctic," Proceedings 17th International Conference on Port and Ocean Engineering under Arctic Conditions (POAC), vol. 1, pp. 203-212, 2003.

[9] M. Johnston and G. Timco, Understanding and Identifying Old Ice in Summer, Ottawa: National Research Council of Canada, 2008.

[10] J. Kulovesi and J. Lehtiranta, "Level Ice Thickness Measurement Using Ship-Based Semi-Automatic Computer Vision," Proceedings of the 22nd IAHR International Symposium on Ice, pp. 111-118, 2014.

[11] M. Garvin, "CCGS Amundsen Field Trials – Spring 2015, Technical Report: OCRE-TR-2016-003," National Research Council, St. John's, 2016.

[12] E. Kim, G. S. Dahiya, S. Loset and R. Skjetne, "Can a computer see what an ice expert sees? Multilabel ice objects classification with convolutional neural networks," Results in Engineering, vol. 4, p. 100036, 2019.

[13] H.-M. Heyn, M. Knoche, Q. Zhang and R. Skjetne, "A system for automated vision-based sea-ice concentration detection and floe-size distribution indication from an icebreaker," in International Conference on Offshore Mechanics and Arctic Engineering, 2017.

[14] M. Rohith, G. Somanath, C. Kambhamettu and C. Geiger, "Stereo analysis of low textured regions with application towards sae-ice reconstruction," in Proceedings of the 2009 International Conference on Image Processing, Computer Vision, & Pattern Recognition, Las Vegas, 2009.

[15] S. Sorensen, V. Veerendraveer, W. Treible, A. W. Mahoney and C. Kambhamettu, "The Polar Sea Ice Topography Reconstruction System," Annals of Glaciology, pp. 1-12, 2020.

[16] M. B. Rivas, J. A. Maslanik, J. G. Sonntag and P. Axelrad, "Sea Ice Roughness From Airborne LIDAR Profiles," IEEE Transactions on Geoscience and Remote Sensing, vol. 44, no. 11, pp. 3032-3037, 2006.

[17] R. Forsberg, K. Keller and S. M. Jacobsen, "Airborne lidar measurements for Cryosat

validation," in IEEE International Geoscience and Remote Sensing Symposium , Toronto, ON, 2002.

[18] A. A. Petty, M. C. Tsamados, N. T. Kurtz, S. L. Farrell, T. Newman, J. P. Harbeck, D. L. Feltham and J. A. Richter-Menge, "Characterizing Arctic sea ice topography using high-resolution IceBridge data," The Cryosphere, vol. 10, pp. 1161-1179, 2016.

[19] N. T. Kurtz, S. L. Farrell, M. Studinger, N. Galin, J. P. Harbeck, R. Lindsay, V. D. Onana, B. Panzer and J. G. Sonntag, "Sea ice thickness, freeboard, and snow depth products from Operation IceBridge airborne data," The Cryosphere, vol. 7, pp. 1035-1056, 2013.

[20] L. Lalumiere, "GPR Capabilities for Ice Thickness," Sensors by Design Ltd., Newmarket, 2011.

[21] J. C. Moore, P. Maijala, S.-E. Hjelt, A. Palli and A. Sinisalo, "GPR investigations of glaciers and sea ice in the Scandinavian Arctic," Natural Resources Institute Finland, 2000.

[22] C. Haas, J. Lobach, S. Hendricks, L. Rabenstein and A. Pfaffling, "Helicopter-borne measurements of sea ice thickness, using a small and lightweight digital EM system," Journal of Applied Geophysics, vol. 67, pp. 234-241, 2009.

[23] C. Haas, J. Lobach, S. Hendricks, L. Rabenstein and A. Pfaffling, "Helicopter-borne measurements of sea ice thickness, using a small and lightweight digital EM system," Journal of Applied Geophysics, vol. 67, pp. 234-241, 2009.

[24] R. J. Galley, M. Trachtenberg, A. Langlois, D. G. Barber and L. Shafai, "Observations of geophysical and dielectric properties and ground penetrating radar," Cold Regions Science and Technology, vol. 57, pp. 29-38, 2009.

[25] H. Liu, K. Takahashi and M. Sato, "Measurement of Dielectric Permittivity and Thickness of Snow and Ice on a Brackish Lagoon Using GPR," IEEE Journal of Selected Topics in Applied Earth Observations and Reomte Sensing, vol. 7, no. 3, pp. 820-827, 2014.

[26] L. Backstrom and H. Eicken, "Capacitance probe measurements of brine volume and bulk salinity in first-year sea ice," Cold Regions Science and Technology, vol. 46, pp. 167-180, 2006.

[27] H. A. Murthy and S. Haykin, "Bayesian Classification of Surface-Based Ice-Radar Images," IEEE Journal of Oceanic Engineering, vol. 12, no. 3, pp. 493-500, 1987.

[28] Rutter Inc., Ice Navigator Detection and Navigation System, St. John's: Rutter Inc..

[29] A. R. Mahoney, H. Eicken, Y. Fukamachi, K. I. Ohshima, D. Simizu, C. Kambhamettu, M. V. Rohith, S. Henricks and J. Jones, "Taking a look at both sides of the ice: comparison of ice thickness and drift speed as observed from moored, airborne and shore-based instruments near Barrow, Alaska," Annals of Glaciology, vol. 56, no. 69, pp. 363-372, 2015.

[30] R. Gagnon, D. Cumming, R. Ritch, R. Browne, M. Johnston, R. Frederking, R. McKenna and F. Ralph, "Overview accompaniment for papers on the bergy bit impact trials," Cold Regions Science and Technology, 2008.

[31] R. Gagnon, "Analysis of data from bergy bit impacts using a novel hull-mounted external Impact Panel," Cold Regions Science and Technology, 2008.

[32] R. Ritch, R. Frederking, M. Johnston, R. Browne and F. Ralph, "Local ice pressures measured on a strain gauge panel during the CCGS Terry Fox bergy bit impact study," Cold Regions Science and Technology, Vol. 52, 2008.

[33] Y. Wang, G. Feng and P. Li, "Research on monitoring point layout of health monitoring system of the icebreaker," Proceedings of the International Conference on Offshore Mechanics and Arctic Engineering, 2017.

[34] G. Wang, K. Pran, G. Sagvolden, G. Havsgård, A. Jensen, G. Johnson and S. Vohra, "Ship hull structure monitoring using fibre optic sensors," Smart Materials & Structures 10, 2001.

[35] A. Bekker, L. Lu, van Zijl, C.M., J. Matthee and P. Kujala, "Correlation between bow ice loads and operational responses during ice navigation in the Weddell Sea," 25th International Conference on Port and Ocean Engineering under Arctic Conditions (POAC), 2019.

[36] M. Johnston, R. Frederking, G. Timco and M. Miles, "MOTAN: A Novel Approach for Determining Ice-Induced Global Loads on Ships," in Proceedings MARI-TECH, 2003.

[37] M. Johnston, "Global ice impact forces on the CCGS Amundsen, Beaufort Sea, Technical Report: CHC-TR-085," National Research Council, Ottawa, 2012.

[38] T. Kouts, M.-J. Lilover and K. Vahter, "On-line system for monitoring and forecast of ship resistance in ice, supporting winter navigation in the Baltic Sea," in IEEE/OES Baltic International Symposium, 2014.

[39] H.-M. Heyn, M. Blanke and R. Skejetne, "Ice Condition Assessment Using Onboard Accelerometers and Statistical Change Detection," IEEE Journal of Oceanic Engineering, pp. 1-17, 2019.

[40] A. Bowen, D. Yoerger, C. German, J. Kinsey, M. Jakuba, D. Gomez-Ibanez, C. Taylor, C. Machado, J. Howland, C. Kaiser, M. Heintz, C. Pontbriand, S. Suman, L. O'Hara, J. Bailey, C. Judge, G. McDonald, L. Whitcomb, C. McFarland and L.

Mayer, "Design of Nereid-UI: A Remotely Operated Unverwater Vehicle of Oceanographic Access Under Ice," in IEEE Oceans, St. John's, 2014.

[41] D. A. Rothrock, Y. Yu and G. A. Maykut, "Thinning of the Arctic Sea-Ice Cover," Geophysical Research Letters, vol. 26, no. 23, pp. 3469-3472, 1999.

[42] A. Bekker, M. Suominen, P. Kujala, R. De Waal and K. Soal, "From data to insight for a polar supply and research vessel," Ship Technology Research, Vol. 66, 2019.

9. AUTHORS BIOGRAPHY

Matthew Garvin holds the current position of Research Engineer / Harsh Environment Marine Operations Team Lead at the National Research Council's Ocean, Coastal, and River Engineering Research Centre in St. John's, NL, Canada. He is responsible for coordinating the research activities of the team with a focus on improving operational efficiency and safety for people working in harsh marine environments while building foundational technologies to support autonomous ships. His previous experience is mainly in private sector engineering consulting to clients in the marine and energy sectors, including an extensive list of test campaigns in the NRC's towing tank, ice tank, and offshore engineering wave basin as well as field work in near-shore and subarctic environments.

REVIEW OF TECHNOLOGIES FOR REGIONAL ICE SEVERITY SENSING

M. Garvin, National Research Council, Canada

SUMMARY

This paper contains a review of technologies employed in the literature to provide data on ice conditions that the ship may encounter in hours to days. Masters use a wide variety of information to assess ice conditions along their planned route. This information is used to assess ice severity and to optimize the route to avoid damage to the ship, besetting, etc. This assessment is made by the officers in charge of the ship based on experience and metocean publications such as weather forecasts and ice charts. A fundamental challenge in allowing autonomy or decision-support for navigation of ice-covered waters is providing ice severity data that feeds decision-making. A framework for data aggregation is proposed, based on available ice sensing technologies. The proposed framework involved aggregating these diverse sets of data into a regional map of forecast ice severity based on a vessel-specific risk assessment such as POLARIS. This would provide a decision-making tool for ship operators and autonomous navigation systems.

1. INTRODUCTION

The work described herein was undertaken as part of a project to specify and evaluate sensor suites for use aboard vessels operating in ice to detect ice severity in real-time. Such a sensor suite is envisioned as an integral part of a decision support system and eventual autonomous ship navigation in ice since assessing ice conditions and integrating that information into routing and navigational decisions is vital to safe and efficient operation. In order to fulfil this objective, the sensor suite must be able to provide both tactical-level information (to allow routing decisions on the scale of hours to days) and operational-level information (to allow navigational decisions on the scale of seconds to minutes). This paper will focus on sources of information that may provide data that is useful at a tactical level. To do this, up-to-date information on the ice severity along a ship's routing options is needed, allowing the master or autonomous autopilot to plot a course that optimizes ice severity and distance travelled to keep risk of damage or besetting within acceptable levels.

This paper is intended for readers who are not familiar with data sources used in ice forecasting, operational ice forecasting products, or arctic shipping risk assessment. The information in this paper will be already well understood by people trained in the relevant disciplines. I hope that the information will provide useful for people outside their regular discipline and potentially provide some insights. A companion paper is being written that investigates operational-level technologies.

2. PROBLEM DESCRIPTION

Navigating in ice-covered waters presents complications for ship navigation. In addition to the risks of structural damage and possible loss of the ship, many ships become beset by ice and require rescue. In some cases, besetting can pose a risk to crew and cargo even if there is no structural damage to the ship, if the duration of the besetting exceeds the supplies aboard the ship. Current practices rely on the training and experience of the ships' masters and an array of forecasting tools to avoid conditions that may result in damage to the ship or besetting.

This paper contains a high-level overview of current tactical-level technologies described in the scientific literature but does not hope to discuss any one of them in technical detail. The goal is to provide enough description to understand the strengths of the various techniques and how they may fit together to provide a complete picture. Many diverse technologies have been used to measure and observe ice for scientific purposes. Taken together, the various techniques promise a formidable battery of ice observations that can assist mariners and eventually autonomous ships in forming an understanding of ice conditions along their route.

In order to estimate the severity of ice confronting a vessel at the tactical level, measurement of the following ice properties is desired: *Thickness, Strength, Concentration, Composition (snow, rubble, brash, ridging, refrozen rubble, 1st year, multi-year, etc.), Piece size (dimensions and mass)*, and *Pressure*.

3. METHODOLOGY

The searches for research literature on real-time ice sensors and/or sensing techniques were undertaken in the following databases*: Scopus, Arctic & Antarctic Regions*, and the *National Technical Information Service*. These search results were used as a starting point; individual publications of interest were identified and many publicly-available sources were referenced in the literature, which were also consulted.

4. REMOTE ICE SENSING TECHNOLOGIES

4.1 SATELLITE-BASED INFORMATION

Satellites provide an opportunity for large area, frequent monitoring of ice conditions and are ideally applicable to making regional ice measurements. A variety of satellite information that has been used in the scientific literature is described below. For satellites currently operating that provide publicly available data, a link to the data has been provided.

There are a number of satellite-based Synthetic Aperture Radar (SAR) systems, including The Canadian Space Agency's RADARSAT series, the European Space

Agency's ERS-2 (mission ended 2019), CryoSat-2, & Sentinel-1. CryoSat-2 carries a SAR, complimented by an interferometric radar altimeter which allows precise measurements of the sensor's height above surface [1]. Sentinel-1 carries a 5.405 GHz c-band SAR with multiple swath widths up to 400 km and resolutions down to 5 m [2]. RADARSAT Constellation is the latest in the range of RADARSAT missions but, launched in June 2019, it has not yet featured in the scientific literature. Its predecessor, RADARSAT-2, carries a 5.405 GHz SAR with a spatial resolution of 8 - 100 m (50 - 500 km swath widths) on a 100.7 min orbit, 8.6 degrees off polar, providing daily near-global coverage [3]. RADARSAT Constellation offers higher resolution, down to 3 m on a 20 km swath. There is significant potential for RADARSAT Constellation's enhanced coverage and expected data on ice extent, density, strength, and age.

In general, SAR excels at providing relatively small-scale, detail information such as floe identification, areas of ridging, and even identification of ice types. RADARSAT-2 and ERS-2 have been used for navigation planning and ice surveillance in near real-time [4]. High resolution images were used to identify ice-covered areas, individual large floes, areas of ridging, and ice age (new ice, nilas, first year ice, multi-year ice) in support of icebreaker-escorted navigation, allowing safer and faster navigation through ice-covered waters [5]. In this case, the interpretation of the images required human involvement in interpretation but the nature of the mission allowed ground-truthing of the information gathered from the satellite images, showing good agreement. Further work has developed a "semi-supervised classification model" in a move towards automated interpretation of SAR data from ERS-2, RADARSAT-2, and Sentinel-1. The semi-supervised method was able to identify the presence of sea ice with good reliability from different SAR data sources [6]. Recent developments have suggested good results are possible when using RADARSAT-2 to classify pure multi-year and first-year ice [7].

The US Defense Meteorological Satellite Program (DMSP) satellites orbit the earth every 101 minutes and provide twice-daily global coverage. While these satellites carry a variety of sensors, the Special Sensor Microwave Imager (SSM/I) is of particular interest. SSM/I is a multi-frequency polarized passive microwave radiometer system that measures microwave brightness temperatures at four frequencies from 19.35 to 85.5 GHz. A more advanced Special Sensor Microwave Image Sounder (SSMIS) has been put into service which increases the highest frequency 91.6 GHz. SSM/I and SSMIS provide differentiation between open water, first year ice, and multi-year ice. This differentiation is based on the emissivity differences between types of ice and open water. From this information, it is possible to determine total ice concentration and concentrations of both first-year and multi-year ice with a resolution of 30 x 30 km. Similar data is collected from Advanced SCATerometer (ASCAT) sensors aboard EUMETSAT MetOp satellites and NASA's Advanced Microwave Scanning Radiometer. Compared to SAR, passive radiometers are better suited to providing large-scale ice information such as the identification of ice extent. [8, 9]

Polar Operational Environmental Satellites (POES), operated by NOAA, provide daily data by making 14 polar orbits daily which, combined with the rotation of the earth provide complete polar coverage. Of particular interest is the Advanced Very High Resolution Radiometer (AVHRR) that allows measurements of surface emissivity with approximately 1 km resolution in the absence of cloud cover. Combined with data from the DMSP SSM/I, POES AVHRR provides complete ice coverage data for ice-covered navigable waters. [10, 8]

Many satellite missions also carry visual-spectrum cameras. CCD near-visible spectrum camera data (0.43 – 0.90 μm) from China's HJ-1A/B has been used to monitor sea ice coverage and extent based on comparisons of reflectance between snow, sea ice, and open water but has not successfully been able to distinguish multi-year ice [11]. Similar capabilities are offered by the MODerate-resolution Imaging Spectro-radiometer (MODIS) instruments on NASA's Earth Observing satellites.

Satellite data is widely used by meteorologists but does have limitations. Atmospheric conditions can obscure measurements and sea temperature variations such as pooled melt water can result in misleading data. This potential ambiguity is a barrier to the widespread operational use of satellite data. To help make satellite data more complete and useful, techniques are available to interpolate gaps in the data. [12]

Work has been done to use satellite-based SAR and LIDAR data to estimate ice thickness. This is a very enticing prospect due to the large areas that can be covered. Ice, Cloud, and Land Elevation Satellite-2 (ICESat-2), following on from its predecessor ICESat, carries a laser-based precision height measurement instrument that can be used to measure the height and profile of ice sheets, thereby providing an estimate of thickness. Current methods rely on assumptions about ice density to translate freeboard into ice thickness (having difficulty distinguishing between ice and snow-cover, for example) and are difficult to ground-truth [13]. Satellite-based LIDAR from ICESat has been used to measure the freeboard of sea ice with uncertainties of 7 cm using "tiepoints" of open water to provide a sea-surface height reference to which to compare the ice surface height [14]. ICESat data has been combined with AMSR-E data (a similar technology to SSM/I that is no longer in service) to measure sea ice thickness with an uncertainty of up to ~58 cm [15]. This large uncertainty is attributed to measurement uncertainties of the two datasets and assumptions about snow and ice densities. The US National Snow & Ice Data Center provides a dataset of sea ice freeboard and thickness based on combined data from GLAS and the SSM/I but it is expected that the uncertainty in this data is quite high.

4.2 AERIAL INFORMATION

Aerial ice observation has a wide range of complexity. Even simple visual fly-overs can provide valuable information on the presence and location of leads, floe size, existence of areas of ridging, and information on the degradation and age of the ice (eg. by observation of ponding, drainage, floe shape, etc.) [13]. A trained observer can discern significant information from aerial photographs that is relevant to navigation and this type of observation is still used, particularly when conditions are not favourable for instrument-based observations. In terms of application, aerial observation is best suited for investigating specific areas of interest to be combined with larger-area satellite based data. Heavily-trafficked near-shore areas are also likely to be investigated using aerial observation which is often used to complement shipboard observations; most Canadian Coast Guard ships operating in ice can access data collected from aircraft to be interpreted by on-board ice specialists [16].

Aerial-based Synthetic Aperture Rader (SAR) has been used in a number of studies and has been shown to be capable of providing a wide range of high-resolution (on the order of 15 m) ice information including ice types (first-year, multi-year, Nilas, etc.), floe size, and concentration [17, 8]. While excellent information can be collected from aerial SAR, the spatial and temporal coverage of such techniques is limited to the swatch of ocean surface visible from the aircraft as it flies over and the flight endurance of the aircraft. There is potential for this type of technology to be deployed aboard unmanned drones to increase the area that can be covered but this technology is best suited to areas of specific interest, relatively near aircraft infrastructure.

Applications of aircraft-borne LIght Detection And Ranging (LIDAR), Ground-Penetrating Radar (GPR) and Electromagnetic Induction (EM) sensors are discussed in the companion paper.

4.3 LAND & SEABED-BASED INFORMATION

In near-shore environments, ground-based radar systems have the ability to track icebergs over significant areas. Khan et. al. demonstrated that a high frequency surface wave radar system can track objects including icebergs and sea ice over an area of 160,000 square kilometres [18]. Shore-mounted X-band radar can provide sea ice observation with a shorter range (ice presence, movement, and ridging identification). This type of technology is mature and particularly applicable for shipping routes that pass near shore-based infrastructure such as the St. Lawrence Seaway and the Grand Banks of Newfoundland.

A method of measuring ice which has become increasingly popular is the use of moored, upward-looking Ice Profiling Sonars (IPS) or Upward-Looking Sonar (ULS) that measure ice draft as it passes overhead. This method provides a more accurate thickness estimate than above-surface techniques since it is not sensitive to the presence of snow, although accuracy can be affected by the presence of melt water on the ice surface. IPS is often deployed with an Acoustic Current Doppler Profiler (ADCP) to estimate velocity of the ice passing overhead which is a critical measurement in estimating draft from IPS data. With care in the interpretation of the data, including correct estimates of depth-averaged speed of sound, correction of tilt and depth, and beam footprint effects, draft measurement precision of ±0.1 m or better can be achieved. [19, 20, 21]

For level ice, this allows an estimate of thickness in a similar manner to estimating thickness from freeboard and is subject to similar assumptions regarding ice density. Accuracy can be improved if the sensor is periodically presented with regions of open water which provide a confirmation of the "zero" reading. Absent periodic measurements of open water regions or another means of verifying the signal offset offsets may be present in the data which is particularly problematic in areas with consistent winter ice cover; fall and spring periods of open water bounding the time period of continual ice cover can be used in post-processing data but are less useful for real-time measurement. Roughness of the bottom surface of the ice can also be measured using ADCP and is less impacted by signal drift making ADCPs suitable for detecting areas of ridging within level ice. Sensors combining IPS and ADCP are particularly suited to monitoring ice thickness, ridge keels, and ice movement in frequently travelled ice-covered shipping lanes. [19, 22, 23]

4.4 FORECAST TOOLS & INFORMATION

Ice Charts are issued by a number of forecasting agencies around the world. In North America, Ice Charts are issued by the Canadian Ice Service and the US National Ice Center for the ice-covered waters around the northern coasts of North America. A description of how the NIC incorporates Satellite, Aerial, and ground-truthing into their ice forecasts is provided in [12] but suffice to say that the agencies responsible for producing Ice Charts are active in incorporating the most current available techniques into their forecasts. As such, Ice Charts provide excellent overviews of ice conditions that a ship may encounter and are consequently invaluable tactical route-planning tools. Specifically, products such as satellite SAR imagery are used directly and in near real-time by members of the CIS aboard Canadian Coast Guard ships [24]. It is important to note, however, that Ice Charts are intended to provide regional forecasts and local conditions may differ; yielding the requirement for on-ship observations.

Ice Charts define regions of roughly consistent ice coverage according to: Total Concentration, Partial Concentration of different ice types present (shown according to their thickness), Stage of Development (eg. New ice, first-year ice, multi-year ice, etc.) of the ice types present, and the form (eg. brash ice, small floe, large floe, fast ice, etc.) of the ice types present. This information is conveyed by one or a combination of different methods, depending on the agency, including the "Ice Egg", symbols, and colour-coding overlaid on navigational charts. [25]

The NRC pressured ice model is one of a number of models that forecasts ice deformation, ridging and pressure (compression). The model output provides information to ship captains and operators on the development of compressive ice along shipping routes as well as the drift of pack ice. The model may also be used to quantify parameters that lead to vessel besetting and assessing risk of vessel besetting. The model provides up to a 48-hour forecast with integrated risk assessment under AIRSS & POLARIS (discussed below). At this time, longer-term forecasts are not possible, as the input data for wind and ocean current information is not available on a time frame that would allow this type of longer-term forecast to be accurate. Ice and environmental inputs are provided by the Canadian Ice Service, the Canadian Meteorological Centre and the Department of Fisheries and Oceans. These standard input models have a number of limitations. Some considerations with respect to the pressured ice model's output include grid cell size limitations, initial ice information, wind and ocean current forecast model availability and accuracy constraints. The use of standard ice chart forecasts without accounting for localized conditions will lead to discrepancies in model forecasts [26, 27].

4.5 HISTORICAL DATA

There are publicly available datasets published which list historical ice properties and can be used to provide a probability-based estimate of future ice conditions at long time scales. This type of method is particularly strong at estimating milestone events such as ice formation and breakup but can also provide more nuanced data such as ice conditions, ice hazards, and physical properties. The Canadian Ice Service maintains a public-access database of ice thickness & snow depth measurements for 11 arctic locations (195 locations from 1947 to 2002) [28]. Since these datasets can be onerous to interpret, database tools that compile this data with other data relevant to mariners have proven useful. The Canadian Arctic Shipping Risk Assessment System (CASRAS) pulls together historical datasets including "*marine weather, marine hydrography, physical oceanography, sea-ice conditions, ice hazards and physical properties, Marine Protected Areas, community information, and mariner knowledge including notes, charts, and digital media* " [29]. CASRAS includes a route planning tool that allows masters to access the datasets relevant to their area of operation. Data can be displayed as visualization of historical conditions, reports on specific hazards, and as chart overlays where an intended route can be plotted with data of interest (eg. historical ice break-up). CASRAS includes integrated route planning with AIRSS and POLARIS so expected Ice Number and RIO (discussed below) can be calculated and a probability of exceeding allowable ice conditions for the ship can be assessed based on historical conditions along the proposed route. [29]

There has been some recent effort to develop forecasts based in historical data and, while this work has not been published, it provides an insight into a potential method for long-term ice forecasting. The work in question used metocean data from the ERA-Interim global atmospheric reanalysis combined with OSTIA sea surface temperature and sea ice data for specific regions of the Canadian Arctic. 23 years of historical data from these sources was used to train a convolutional neural network to predict sea ice concentrations based on metocean forecasts. The model was tested on 5 years' worth of data and validated against a further 5 years. The results were promising but further work is needed to integrate operational forecasts, forecast larger regions, and improve resolution.

4.6 RISK ASSESSMENT TOOLS

The Arctic Ice Regime Shipping System (AIRSS) was implemented by Transport Canada as a regulatory standard to allow mariners to assess and minimize risk of damage to their vessel and to thereby avoid pollution in the Arctic. AIRSS combines information about the ice severity with information about the vessel's capability to provide a "GO/NO-GO" decision for mariners. Under AIRSS, the Ice Regime (a region of more or less consistent ice conditions) is accounted for by based on Ice Concentration, Stage of Development, Stage of Decay, and Ice Roughness, generalized as a set of Ice Types as observed from the ship itself as a supplement to published ice charts. Similarly, each vessel is assigned a "Vessel Class" ("Canadian Arctic Class" 1 thru 4 and "Type" A thru E in order of descending ice capability) and . Each Vessel Class is assigned a set of numeric "Ice Multiplier" (one for each ice type) based on the displacement, structural strength, and power of the vessel and defined by the regulation. The relative concentration of each Ice Type is combined with the Ice Multiplier for that ice type and the products are summed to determine an "Ice Number". If the Ice Number is greater than zero, navigation is permitted. The regulations further simplify the application of the AIRSS by subdividing the Canadian Arctic waters into sixteen Shipping Safety Control Zones with specified opening and closing dates for ships in each of the Vessel Classes. [30, 31]

The Polar Operational Limit Assessment Risk Indexing System (POLARIS), published by IMO, provides a similar method of risk evaluation to AIRSS. POLARIS is not mandatory or prescriptive, intended rather as a decision support tool. Under POLARIS, ships are assessed based on their ice capability and assigned a Risk Index Value (RIV). Levels of ice capability, in descending order, include "Polar Class" 1 thru 7, "1A Super", "1A", "1B", and "1C", based on the IACS Polar Class and the Finnish-Swedish Ice Class. The ship category is compared to ice conditions defined according to the World Meteorological Organization nomenclature as used for international ice charts, ice regime, and state of decay. The RIV is compared to the ice conditions to compute a Risk Index Outcome (RIO) which is then used to assess risk as "Normal Operation", "Elevated Operational Risk", or "Operation Subject to Special Consideration" (lower RIOs indicating higher risk). The higher two risk assessments may result in restrictions to operation such as limited speed, more rigorous watchkeeping, or icebreaker support.

Rather than a prohibition, ships' masters are given discretion to operate in the highest risk level. [32]

The nature of the Ice Number calculated under AIRSS or the RIO calculated under POLARIS ostensibly provides a numeric scale by which to compare the relative risk of various routing options which are permissible within the regulations with higher AIRSS Ice Numbers representing lower risk.

5. INFORMATION GAPS

Remote sensing of ice thickness is an ongoing but critical challenge since thickness is the most important parameter to assess ice severity for shipping. Current methods using satellite-based measurements rely on significant assumptions, yielding uncertainties that are too high for reliable use in estimating the risk to shipping. Current remote sensing techniques rely on validation with direct measurement at near-shore stations and occasional ship-based direct measurements but lack accuracy.

Measurements of ice extent, piece size, and concentration are quite reliable from satellite and aerial observation, allowing good assessment of these conditions for large regions. Using metocean data as an input, models can predict regional ice pressure with accuracy. Distinguishing between first-year, second-year, and multi-year ice is an area that needs attention since the strength of ice varies significantly between these types and is the second most important parameter to assess ice severity from the point of view of ship navigation. In areas where multi-year ice is a potential problem, it is especially important to be able to identify pieces of multi-year ice within regions of first-year oce cover; the harder, stronger multi-year ice presents a hidden threat. Currently, differentiation relies on differences in colour and surface topography (rubbling vs. ridging vs. smooth rounding) [16]. There are still questions about the reliability of satellite imagery in differentiating first-, second-, and multi-year ice [13]. Currently human examination of the satellite data is needed but tools may be developed that allow automation.

6. PROPOSED APPROACH

For the current, near-term, and long-term timeframes, the information should be presented in a format that allows assessment of different routing options in a way that can be interpreted by the end user. At its most basic, a risk assessment could be based on a simple numeric scale such as AIRSS or POLARIS which would allow a comparison of options based on the expected ice conditions along the route and the capability of the vessel. Risk data overlaid on a marine chart for interpretation by mariners is in-line with current ice chart formats and can be converted into a set of regions of ranked risk based on a scaled metric such as the POLARIS RIV or AIRSS Ice Number, similar to that shown in Figure 1, from Kubat et. al. [29]. In Figure 1, alternate route A (dashed line) minimizes the distance sailed in heavy ice while alternate route B avoids areas of higher risk (as indicated by AIRSS Ice Numeral). We may expect route A to be suitable for a ship that is structurally capable of navigating severe ice but that has a severe speed penalty associated with severe ice while rout B would be suitable for a ship not structurally capable of navigating severe ice. Combined with information on the ship's destination and schedule requirements, this type of model would allow identification of recommended routes, compare the ice risk between routing options, and identify potential schedule variances. This trade-off would require vessel-specific optimization.

In order to provide up-to-data information on current conditions and near- and long-term forecasts, a cloud-based data storage platform is needed where data from the various sources is maintained and updated. Multiple ships providing observations of their local conditions would provide verification of forecasts with a larger number of contributing vessels strengthening the dataset. A proposed framework that includes aggregating data from multiple ships and other sources to be optimized against schedule requirements and the ship-specific risk assessment is shown in Figure 2.

This approach is not entirely novel. Eriksson et. al. [33] incorporated a range of data types including weather modelling and satellite observations to optimize ship routing based on observed and forecast surface current and wind data. A system with similar goals and principles has been described by Zhang et. al. [34]. More recently, a holistic risk assessment framework has been proposed by Browne et al. [35] that would allow risk to be assessed based on impacts to life safety, ecology, and socioeconomics rather than simply vessel-centric risk. This approach would account for the very different life safety risks posed by an accident in a remote location involving a ship with a high number of persons on board compared to a ship with few persons on board. Similarly, the ecological impacts of an accident involving a ship with dangerous cargo would involve a different risk calculation than a ship with less potential to cause ecological damage. Under this type of framework, different types of ships could be routed differently given their different risk profiles (for instance, a ship with dangerous cargo would be routed further from ecologically sensitive areas and ships with high numbers of persons on board would be routed closer to areas with high search and rescue capacity).

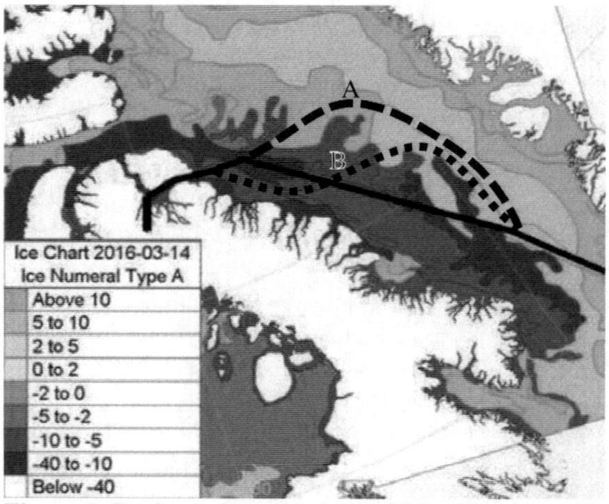

Figure 1. Regions of expected AIRSS Ice Number with alternate routes overlaid, allowing along-route risk assessment. From Kubat et. al. [29], with alternate (dashed) routes added.

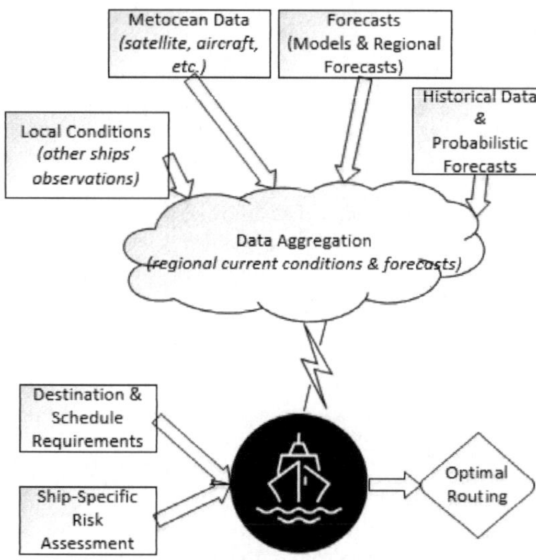

Figure 2. Proposed tactical information model.

7. CONCLUSIONS

There is a wealth of ice data available from various sources. In order to make this data more useful for informing navigational decisions, a significant data aggregation / data fusion effort is needed. The goal of this task is to pull data from the various sources and make it available in a format that provides a current, near-term forecast, and long-term forecast picture of ice conditions. This would allow ships' captains to assess the likely ice conditions along their route and adjust their route and speed accordingly. While it is acknowledged that there is significant nuance involved in aggregating diverse data sets and significant technical challenges associated with even compiling and storing this data, the following general principle is proposed: a snapshot of "current conditions" would be based on an aggregations of the most recent (last ~12 hours) data from satellites, aircraft flyovers of areas of particular interest, and shipboard observations. Currently-produced ice charts and ice pressure models typically provide several days' worth of forecast data at a regional level, and beyond the time window of current forecast models, tools built on historical data can provide probabilistic insight into what conditions may be expected. There are significant historical datasets available for a wide variety of ice characteristics. These have great potential as tools to develop and validate probabilistic forecasting tools and could form a backbone of data aggregation efforts.

Modern communications technologies are advancing at an unprecedented pace but there are still challenges associated with providing high-bandwidth, reliable data transfer to ships navigating around the globe. If ships are to rely on data from outside sources for navigational decision making, the delivery of this data must be very robust.

8. REFERENCES

[1] European Space Agency, "CryoSat-2 operations," European Space Agency, [Online]. Available: https://www.esa.int/Enabling_Support/Operations/CryoSat-2_operations. [Accessed 30 04 2020].

[2] European Space Agency, "Sentinel-1," European Space Agency, [Online]. Available: https://sentinel.esa.int/web/sentinel/missions/sentinel-1. [Accessed 30 04 2020].

[3] Canadian Space Agency, "RADARSAT satellites: Technical comparison," Canadian Space Agency, 19 12 2019. [Online]. Available: https://asc-csa.gc.ca/eng/satellites/radarsat/technical-features/radarsat-comparison.asp. [Accessed 18 04 2020].

[4] P. W. Vachon, P. Adlakha, H. Edel, M. Henschel, B. Ramsay, D. Flett, M. Rey, G. Staples and S. Thomas, "Canadian Progress Toward Marine and CoastalApplications of Synthetic Aperture Radar," JOHNS HOPKINS APL TECHNICAL DIGEST, vol. 21, no. 1, pp. 33-40, 2000.

[5] L. H. Pettersson, S. Sandven, O. Dalen, V. Y. Alexandrov, V. V. Melentyev, V. A. Bogdanov and N. I. Babich, "Satellite radar ice monitoring for winter navigation in the Kara Sea," European Space Agency, 2000.

[6] A. M. Johansson, E. Malnes, S. Gerland, A. Cristea, A. P. Doulgeris, D. V. Divine, O. Pavlova and T. R. Lauknes, "Consistent ice and open water classification combining historical synthetic aperture radar satellite images from ERS-1/2, Envisat ASAR, RADARSAT-2 and Sentinel-1A/B.," Annals of Glaciology, pp. 1-11, 2020.

[7] A. Komarov and M. Buehner, "Detection of First-Year and Multi-Year Sea Ice from Dual-Polarization SAR Images Under Cold Conditions," IEEE

Transactions on Geoscience and Remote Sensing, vol. 57, no. 11, pp. 9109-9123, 2019.

[8] S. Sandven and O. M. Johannessen, "The use of microvave remote sensing for sea ice studies in the Barents Sea," ISPRS Journal of Photogrametry and Remote Sensing, vol. 48, no. 1, pp. 2-18, 1993.

[9] NOAA Office of Satellite and Product Operations, "Defense Meteorological Satellite Program (DMSP)," 13 06 2018. [Online]. Available: https://www.ospo.noaa.gov/Operations/DMSP/index.html. [Accessed 31 03 2020].

[10] NOAA Office of Satellite and Product Operations, "Polar Operational Environmental Satellites (POES)," 27 01 2017. [Online]. Available: https://www.ospo.noaa.gov/Operations/POES/index.html. [Accessed 31 03 2020].

[11] G. Zheng, J. Bai, S. Zhang, B. Zhu, J. Sun, F. Wu and C. Wang, "Bohai bay sea ice monitoring based on the HJ satellite images," in 7th International Congress on Image and Signal Processing, Dalian, China, 2014.

[12] K. C. Partington, "A Data Fusion Algorithm for Mapping Sea-Ice Concentrations from Special Sensor Microwave/Imager Data," IEEE Transactions on Geoscience and Remote Sensing, vol. 38, no. 4, pp. 1947-1958, 2000.

[13] M. Johnston and G. Timco, Understanding and Identifying Old Ice in Summer, Ottawa: National Research Council of Canada, 2008.

[14] R. Kwok, G. F. Cunningham, H. J. Zwally and D. Yi, "Ice, Cloud, and land Elevation Satellite (ICESat) over Arctic sea ice: Retrieval of freeboard," Journal of Geophysical Research, vol. 112, no. C12013, 2007.

[15] D. Yi, H. J. Zwally and J. W. Robbins, "ICESat observations of seasonal and interannual variations of sea-ice freeboard and estimated thickness in the Weddell Sea, Antarctica (2003-2009)," Annals of Glaciology, vol. 52, no. 57, pp. 43-51, 2011.

[16] Canadian Ice Service, "Manual of Ice (MANICE) Chapter 2: Ice Observations," Environment and Climate Change Canada, 06 2005. [Online]. Available: https://www.canada.ca/en/environment-climate-change/services/weather-manuals-documentation/manice-manual-of-ice/chapter-2.html. [Accessed 05 05 2020].

[17] R. A. Schuchman, L. L. Sutherland, O. M. Johannessen and E. D. Leavitt, "Geophysical Information On The Winter Marginal Ice Zone Obtained From SAR," Geoscience and Remote Sensing Symposium, vol. 2, 1988.

[18] R. Khan, B. Gamberg, D. Power, J. Walsh, B. Dawe, W. Pearson and D. Millan, "Target detection and tracking with a high frequency ground wave radar," IEEE Journal of Oceanic Engineering, vol. 19, no. 4, pp. 540-548, 1994.

[19] A. Shcherbina, D. Rudnick and L. Talley, "Ice-Draft Profiling from Bottom-Mounted ADCP Data," Journal of Atmospheric and Oceanic Technology, vol. 22, pp. 1249-1266, 2005.

[20] A. R. Mahoney, H. Eicken, Y. Fukamachi, K. I. Ohshima, D. Simizu, C. Kambhamettu, M. V. Rohith, S. Henricks and J. Jones, "Taking a look at both sides of the ice: comparison of ice thickness and drift speed as observed from moored, airborne and shore-based instruments near Barrow, Alaska," Annals of Glaciology, vol. 56, no. 69, pp. 363-372, 2015.

[21] E. Ross, D. Fissel, T. Mudge, A. Kanwar and D. Sadowy, "The Sensitivity of Ice Keel Statistics to Upward Looking Sonar Ice Draft Processing Methods," in OTC Arctic Technology Conference, Houston, 2014.

[22] V. Strass, "Measuring sea ice draft and coverage with moored upward looking sonars," Deep-Sea Research, vol. 45, pp. 795-818, 1998.

[23] R. Birch, D. B. Fissel, K. Schaudt, H. Melling, K. Vaudrey, J. Heideman and W. Lamb, "Ice-profiling sonar," Sea Technology, vol. 41, no. 8, 2000.

[24] B. Ramsay, M. Manore, L. Weir, K. Wilson and D. Bradley, "Use of radarsat data in the canadian ice service," Canadian Journal of Remote Sensing, vol. 24, no. 1, pp. 36-42, 1998.

[25] Canadian Ice Service, "Interpreting ice charts," Government of Canada, 07 03 2016. [Online]. Available: https://www.canada.ca/en/environment-climate-change/services/ice-forecasts-observations/publications/interpreting-charts.html. [Accessed 10 04 2020].

[26] I. Kubat, D. Watson and M. and Sayed, "Characterization of Pressured Ice Threat to Shipping," in Proceedings of the 21st International Conference on Port and Ocean Engineering under Arctic Conditions, POAC, 2011.

[27] I. Kubat, M. Sayed, S. Savage and T. Carrieres, "The Response of Sea Ice Dynamics Models to Wind," in Proceedings of the International Conference on Port and Ocean Engineering iunder Arctic Conditions, Lulea, Sweden, 2009.

[28] Environment and Climate Change Canada, "Ice Thickness Program Collection," Open Government Canada, 25 09 2016. [Online]. Available: https://open.canada.ca/data/en/dataset/9e63fd96-65d0-4e5e-817f-c400280cb319. [Accessed 04 05 2020].

[29] I. Kubat, L. Charlebois, R. Burcher, P. Lamontagne and D. Watson, "Canadian Arctic Shipping Risk Assessment System," in Proceedings of the 24th

International Conference on Port and Ocean Engineering under Arctic Conditions, Busan, 2017.

[30] G. Timco and M. Johnston, Arctic Ice Regime Shipping System - Pictoral Guide, Ottawa: National Research Council of Canada, 2003.

[31] Transport Canada, "Acts and Regulations," Government of Canada, 20 01 2010. [Online]. Available: https://www.tc.gc.ca/eng/marinesafety/debs-arctic-acts-regulations-menu-2272.htm. [Accessed 10 04 2020].

[32] L. Fedi, L. Etienne, O. Faury, P. Rigot-Muller, S. Stephenson and A. Cheaitou, "ARCTIC NAVIGATION: STAKES, BENEFITS AND LIMITS OF THE POLARIS SYSTEM," Journal of OCean Technology, vol. 13, no. 4, pp. 58-71, 2018.

[33] L. E. B. Eriksson, Y. Ye, L. Jonasson, W. Qazi, W. Mao, H. Wang, J. Moller, K. Lemmens and S. Dokken, "EONAV – COPERNICUS DATA IN SUPPORT OF MARITIME ROUTE OPTIMIZATION," in IEEE International Geoscience and Remote Sensing Symposium, Valencia, Spain, 2018.

[34] Z. Zhang, D. Huisingh and M. Song, "Exploitation of trans-Arctic maritime transportation," Journal of Cleaner Production, vol. 212, pp. 960-973, 2019.

[35] T. Browne, R. Taylor, B. Veitch, P. Kujala, F. Khan and D. Smith, "A Framework for Integrating Life-Safety and Environmental Consequences into Conventional Arctic Shipping Risk Models," Applied Sciences, vol. 10, no. 8, p. 2937, 2020.

9. AUTHORS BIOGRAPHY

Matthew Garvin holds the current position of Research Engineer / Marine Operations Autonomy & Safety Team Lead at the National Research Council's Ocean, Coastal, and River Engineering Research Centre in St. John's, NL, Canada. He is responsible for coordinating the research activities of the team with a focus on improving operational efficiency and safety for people working in harsh marine environments while building foundational technologies to support autonomous ships. His previous experience is mainly in private sector engineering consulting to clients in the marine and energy sectors, including an extensive list of test campaigns in the NRC's towing tank, ice tank, and offshore engineering wave basin as well as field work in near-shore and subarctic environments.

AN INNOVATIVE MACHINE LEARNING SYSTEM FOR REAL TIME CONDITION MONITORING OF SHIP MACHINERY

S Oikonomou, **I. Lazakis**, University of Strathclyde, UK
G. Papadakis, AP&A Group, UK

SUMMARY

In this paper an innovative on-line condition monitoring system is introduced. It consists of an object-oriented database, a machine learning algorithm and a model to predict machinery failure. The in-house built Object-Oriented Condition Monitoring Database (CMD), avoids the challenges of relational-object mismatch issues which are common when using relational databases in complex applications involving machine learning techniques. The database intelligently stores data from various sensors and then feeds the data into a pipeline to the diagnostic and prognostic system, offering a constant evaluation of the ship machinery at high speed and accuracy. The suggested Condition Based Maintenance (CBM) framework is based on detecting the change of the condition of the machinery in real time by utilizing the Local Outlier Factor (LOF) algorithm for novelty detection. Two case studies are presented that include real-life data from sensors onboard a tanker ship and prediction of failure of the cylinders of two Diesel Generators (DGs).

1 Introduction

In the marine sector data are usually being used for manually monitoring alarms and trends on-board with some cases of simple automated data diagnostic and prognostic procedures [1]. These simple diagnostic and prognostic systems combined with expert knowledge are part of the first generation of Condition Based Maintenance (CBM) systems. As the name states, CBM refers to the process of planning maintenance of a machinery not through fixed time intervals (hard maintenance) but according to the actual condition of the machinery. Condition monitoring has as a goal to collect data regarding machinery condition, called condition monitoring data [2]. Statistical analysis of condition monitoring data is used by most of the existing commercial CBM systems in shipping and is meant to draw population inferences from a sample and create a mathematical model of the data generation process [3]. However, failure mechanisms of machinery have been proven not so easy to model since they are mostly random in nature [4],[5],[6] and [7]. That is why the research community has turned to machine learning techniques in order to solve the problem of unplanned maintenance (e.g. [8]; [9]; [10]). Machine Learning (ML) is the science of programming computers so they can learn from data [11]. Machine learning algorithms could be trained with data, gathered mainly from sensors, and then used for prediction of failure of the machinery.

The real problem that Data Scientists are facing is that most data-driven techniques proposed for CBM in shipping require large amount of historical data, including failure rate data, which are notoriously difficult to find. In addition, a couple of factors make a commercial CBM system an almost impossible task. The main factors are: The existence of different operational profiles of the ship (port maneuvering, bad weather, cruising, etc.), the need for large amount of computation power for training of the algorithms and the mismatch between relational databases and the algorithms. This can also be seen by the fact that only 2% of the total percentage of classed ships have condition monitoring scheme in place [12].

In this paper a novel CBM framework will be presented that will be able to perform a fast and reliable detection of the degradation of a machinery through the use of the CMD and the LOF algorithm and the results will then be streamlined to the diagnostic and prognostic system. The proposed CBM framework will require minimum amount of memory, no historical data and no need for prior human knowledge (experts).

Section 2 presents an overview of the existing CBM approaches and the challenges of their applications in the marine sector. Section 3 gives an overview of the purposed CBM framework. Section 4 presents the set-up of the case studies and results. Section 5 gives a conclusion and section 6 the proposed future research. Section 7 is the acknowledgments section.

2 Condition Based Maintenance

2.1 CBM Approaches for Diagnosis and Prognosis

In the current literature, there are two major CBM approaches namely, the model driven approach and the data driven approach ([13]; [14]; [15]).

The model driven approach incorporates a physical understanding of the targeted system [14] and includes techniques such as rule based expert systems, finite-state machines and qualitative reasoning or else fuzzy modelling ([15]; [16]). As the name states, expert systems require significant human expertise in order to find a solution. The main drawback of these systems is the inability to generalize (they are specific to one system), to adapt in new conditions and to learn from their mistakes ([15]; [14]). In addition, in qualitative reasoning, based on cause/effect analysis, the fault detection is accomplished by comparing the predicted behaviour of a system with the actual observation [15]. That means that by simulating qualitative models of normal and faulty equipment someone can describe the different possibilities of behaviour during operation of a system without the need of numerical models. Sensor data from actual processes are used to select between the different models [15].

The data driven approach is also known as the machine learning approach, which uses historical data to automatically learn of the system behaviour [17]. These models support the recognition and classification of complex fault patterns without much knowledge about the process, the signals and the fault patterns [18]. Some popular machine learning techniques that can be used for the data driven approach are Neural Networks (NN), Decision Trees (DT), and Support Vector Machines (SVM).

The above approaches however present some challenges in their application in the marine sector. The model driven approaches have provided limited solutions for only specific machinery in specific operational conditions and they fail to generalize and to represent with accuracy time-varying and spatially varying phenomena [15]. On the other hand, the data driven approaches require a large amount of quality labelled training data for every case examined, including data from different operational profiles of the ship and/or failure rates. These data require a substantial amount of time to be gathered, a large amount of storage space on-board, database management systems that can handle large amount of data, increased costs to transmit data onshore via satellite for analysis and additional time for training algorithms and having results.

3 New Generation CBM for the Marine Sector

Through the critical review performed on the current literature, it has been shown that existing CBM frameworks have limited applicability and are mostly based on simple statistical analysis. In this respect, it is more than ever necessary to propose a CBM framework that will incorporate advanced machine learning techniques and modelling in order to offer better diagnostic and prognostic abilities than the existing systems with also an increased practical value.

3.1 Proposed CBM framework

In order to deal with the main problems for implementing CBM in shipping, an innovate approach is followed. The new CBM framework has a simple architecture and can be applicable in more than one machinery systems without the need of any historical data and more importantly the intervention of any human (expert). To do this, a database that will facilitate the machine learning process with its object-oriented nature is developed as well. The database will intelligently store data only for condition monitoring and clear the memory every time the training of the algorithm has been finished. Then, a data driven approach will be used to assess the degradation of the machinery and streamline the results to a failure model to predict maintenance.

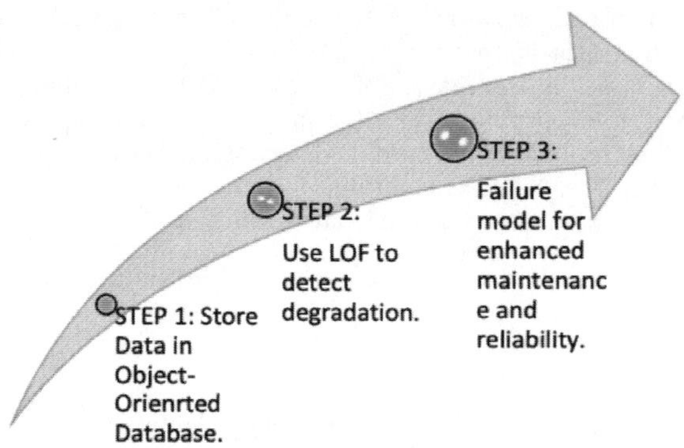

Fig. 1. Purposed CBM framework.

The characteristics of the new CBM framework (see Fig. 1) can be summarized as: A simple and effective database for storing condition monitoring data, the LOF algorithm to use the stored data to detect degradation and a failure model to use the results from the machine learning analysis to provide enhanced maintenance and reliability.

In this paper the CBM framework will be applied for predicting failure of a DG of a tanker ship.

3.2 The Zope Object Database

The new database was built entirely in the Python language using the tools provided by the Zope Object Database (ZODB). The Zope Object Database is an open source OOD for transparently and persistently storing Python objects. Because the ZODB is entirely written with Python, no separate language is needed for database operations and there is very little impact on the code to make objects persistent. In addition, there is no database mapper that partially hides the database.

Using an OOD is better than using object-relational mapping since there is almost no seam between code and database and relationships between objects are handled very naturally, supporting complex object graphs without joins [19].

3.3 Object-Relational Impedance Mismatch

Up until now, the Relational Database has been the most commonly used data model and the vast majority of current database systems are based on this model ([20],[21] and [22]). The advent of the domination of

Object-Oriented programming in software development and the urge to overcome some limitations of the relational model led to the development of the Object-Oriented Databases [20],[22].

For the proposed CBM framework, a database is used in conjunction with Object-Oriented programming (Python). There is a commonly known problem between Relational Databases and Object-Oriented programming called Object-Relational Impedance Mismatch (Ireland & Bowers, 2015). This mismatch can occur only when an Object-Oriented program uses a Relational Database for persistence [23]. Instead of managing relations using different tables with common primary keys, the ZODB let developers use normal Python object references. Thus, the ZODB doesn't require a pre-defined structure of columns and data types for the objects it stores, which means that object attributes can easily change both in quantity and type [19]. This can often be a lot harder when using a Relational Database for storage.

From all the above, it is more efficient to use an OOD in the new CBM framework since Object-Oriented Databases are designed so they can be directly or effortlessly integrated with software that is developed using Object-Oriented programming languages [21].

The purpose of the database in the proposed CBM framework is to maximize the speed of the machine learning techniques and not to store thousands of unnecessary data. To do that it intelligently stores data for training and erases them after they have been used to train the algorithm, making space for new training data when they are needed. That is why the new database was named: Condition Monitoring Database (CMD).

3.4 Local Outlier Factor (LOF) algorithm

The LOF is an algorithm first presented by [25] and its purpose was to identify local outliers in a set of data. The algorithm doesn't just categorize the data in a binary way namely, outliers or not but rather assigns to each data point a local outlier factor (LOF) score which shows the degree of a data point being an outlier, something similar to density-based clustering [26]. Recently, the LOF algorithm has start being used also for novelty detection. LOF creates a decision frontier from the decision function learned from a training dataset which consists of 'normal' data and then assess if a new data-point is a novelty or not [27]. The desired decision boundary can be obtained by varying the LOF threshold [28]. LOF works very well in many cases, while often outperforming the competitors, for example in network intrusion detection [21] and on processed classification benchmark data [22].

To conclude, the great general applicability as well as the demonstrated effectiveness compared to other algorithms of the LOF algorithm are the reasons why it has been chosen as the algorithm to be used in our CBM framework.

3.5 P-F Diagrams

Potential-Failure (P-F) diagrams illustrate how a condition deteriorates until it reaches a point of functional failure [29]. Figure 2 shows how most of the CBM systems currently in place work. They use the 'inspection' interval logic where the system is periodically checked to determine the process of the failure. Continuous monitoring systems require large storage abilities to store condition monitoring data plus increased costs to transmit the data onshore via satellite for analysis which makes it unappealing to the operators.

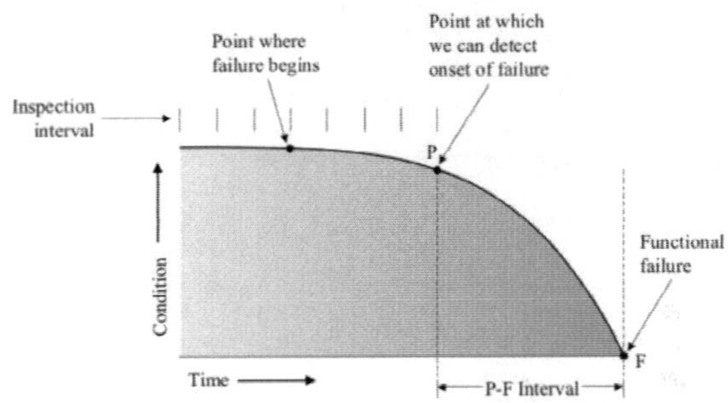

Fig. 2. A Potential-Failure diagram.

These fundamental problems of the shipping industry are usually overlooked by the research community which until now propose systems that are over expensive to implement or simply nonpractical for the complex reality of a ship.

Having into consideration the problems of the current CBM systems a new theory can be made working around these problems. Figure 1 showed an illustration of how the degradation process occurs. This paper suggests taking that intuitive graph and mathematically solve it in order to acquire a general model of the failure mechanism. Since the CBM system can only work from the point where the failure, can be detected and assuming the condition practically remains the same from the point the machinery started working till the point failure can be detected (P), Figure 2 is transformed to Figure 3. The shape of the general P-F interval can now be modelled as an ellipse starting at the point P, which is assumed to be the time where the machinery has maximum reliability (100%). The ellipse will be subject to change according to the deteriorations detected by the LOF algorithm.

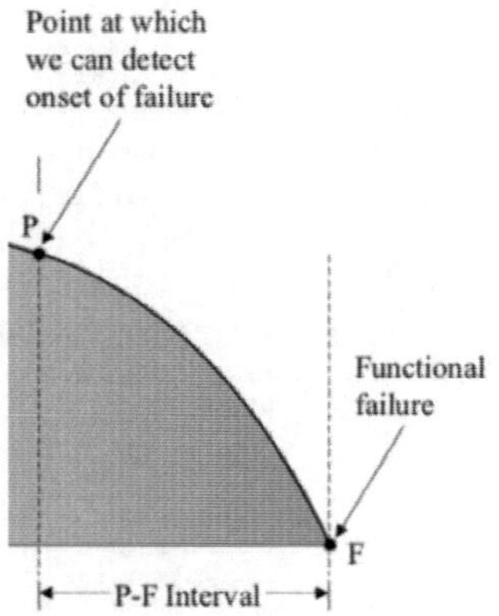

Fig. 3. P-F Graph starting from the point that the failure is detectable.

4 Application of CBM

4.1 Real World Case Studies

In this chapter, the purposed CBM system will be used together with real world data. The dataset that was used has been taken by a tanker and for the period of 3 months, namely: 03/04/2018 to 02/07/2018.

First of all, in order to understand deterioration, one needs to identify the stable operational profiles of the ship. In manuals of the engine of the ship and especially for bridge manoeuvring systems it is shown a ship's engine speed is controlled to be within a range of the RPM that have been set from the bridge. That means that a stable speed operational profile can be identified by when the engine speed is stable inside a predefined range. Even though a stable operational profile requires a stable engine speed it also requires a stable power. The reason is that, for example, a ship can have the same speed for different payloads, but the power required is different and the operational conditions of the machinery are different. For the proposed CBM system it will be required that the power of the engine also needs to be stable between a predefined range in order to have a stable operational profile.

In addition, the weather conditions can also have an effect in the operational conditions of the machinery. All the available relevant measurements in the dataset will be used, namely: the wind speed, wind direction, wind beaufort, speed over ground and M/E Torque. By keeping the speed over ground and the M/E Torque stable we ensure that currents are not affecting the operational conditions (e.g. by increasing resistance to the ship).

Overall, all the above parameters need to be kept stable in order to have a stable operational profile as it can be seen in Figure 2.

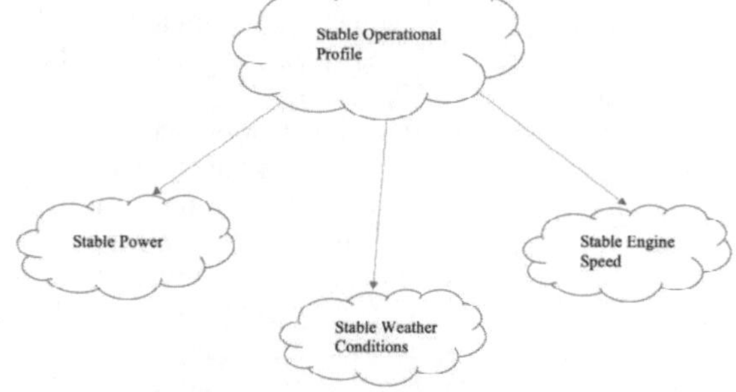

Figure 4: Stable Operational Profiles

4.2 Assumptions

First of all, the reliability (the probability that the machinery will perform its intended function adequately) of the machinery is perfect when it is brand new and equal to 100%) It is assumed that the reliability condition of a machinery after major maintenance is almost perfect.

Furthermore, after discussing with the ship operators it is assumed that the the ship is at sea 80% of the time in a year.

It is assumed that there is a stable operational condition when:

- The measurements from the engine speed and engine power are deviating +-5% as suggested in the bridge manoeuvring systems [30]
- The measurements from the weather dataset (the wind speed, wind direction, wind beaufort, speed over ground and M/E Torque) are deviating +-10%. The +-10% instead of +-5% is introduced due to weather's more volatile nature.

By using these assumptions for a stable operational profile, the dataset was separated into 5 parts, each one of which represents a period that the ship was in stable operational profile.

4.3 DG Cylinder Datasets

The datasets for the Diesel Generator 1 and Diesel Generator 2 cylinders have both as parameters: Exhaust Gases Outlet Temperature, CFW Inlet Pressure, CFW Outlet Temperature, Cooling Air Temperature, LO Inlet Pressure, LO Inlet Temperature and Exhaust Gas Outlet Temperature. The time period of measurements was 3 months, 03/04/2018-02/07/2018 and the time frequency was one minute.

4.4 Maintenance of DG1 and DG2

Because the proposed CBM system takes into account when was the last maintenance procedure for the machinery under evaluation the maintenance log of the machinery onboard the ship was reviewed. Also, because the maintenance logs did not have any overhaul activity for the DG1 cylinders log we will assume that the time of maintenance is the time that the ship overtaken a drydocking (17/06/2017). The data gathering activity started on the 02/04/2018 which is approximately 9 months since overhauling. In addition, after speaking with ship operators and reviewing the manuals of the DGs, the manufacturer suggests maintenance in intervals of 2 years. So, the time that was left until maintenance from the date that data gathering activity started was 1 year and 1 month.

For DG2 the maintenance log said that major overhauling for the cylinders happened on the 11/04/2017 and thus approximately 1 year had passed until when the data gathering activity started and 1 year was remaining for overhauling according to the manufacturer's recommendations.

The months between the overhauling and the data gathering activity are very important to determine the uncertainty of our prediction. That is because the longer the period, the more probable that deterioration has already happened without being monitored and thus compromising the assumption of our model that the machinery is healthy when the data gathering activity started. A sensitivity analysis will be performed for assumed reliability condition of the machinery when the data gathering process started.

4.5 Detected Deteriorations

The total changes of condition (deteriorations) of the DG1 cylinders for the data gathering period: 03/04/2018-02/07/2018 were found to happen on:

- 2018-04-19 09:22:00
- 2018-04-24 17:00:00
- 2018-04-29 04:13:00
- 2018-05-18 22:31:00

For DG2 cylinders the deteriorations for the data gathering period: 03/04/2018-02/07/2018 were found to happen on:

- 2018-04-18 20:37:00
- 2018-04-21 00:30:00
- 2018-04-28 20:43:00
- 2018-05-03 11:30:00

4.6 Sensitivity Analysis

Table 1 shows the sensitivity analysis of the assumed initial reliability for DG1 and DG2 cylinders at the time that data gathering started as well as the time of failure predicted for this reliability by the CBM system.

Table 1: Sensitivity analysis

Index	Reliability	DG1	DG2
A	99.9%	10 yrs-3 m	12yrs-5m
B	90%	2 yrs-4 m	2yrs-10m
C	80%	1 yr-9 m	2yrs-1m
D	70%	1 yr-6 m	1yr-10m
E	60%	1 yr-4 m	1yr-7m
F	50%	1 yr-3 m	1yr-6m
G	40%	1 yr-2 m	1yr-5m
H	30%	1 yr-2 m	1yr-4m
I	20%	1 yr-1 m	1yr-4m
J	10%	1 yr-1 m	1yr-4m

From Table 3 it can be seen that for DG1 as well as DG2, when the reliability of the machinery is assumed to be very high the prediction of the time of failure goes very far into the future. That is logical since it basically means that between the date of previous maintenance and the time of the data gathering (which is 9 months for DG1 and 1 year for DG2) the machinery does not deteriorate at all and so the deterioration process is very slow. In fact, as the assumed reliability of the machinery is reduced the time of failure also decreases.

Figure 3 is a plot for DG1 of the predictions of the model in years remaining until maintenance since the data gathering activity started for different assumptions of the reliability of the machinery when the data gathering activity started.

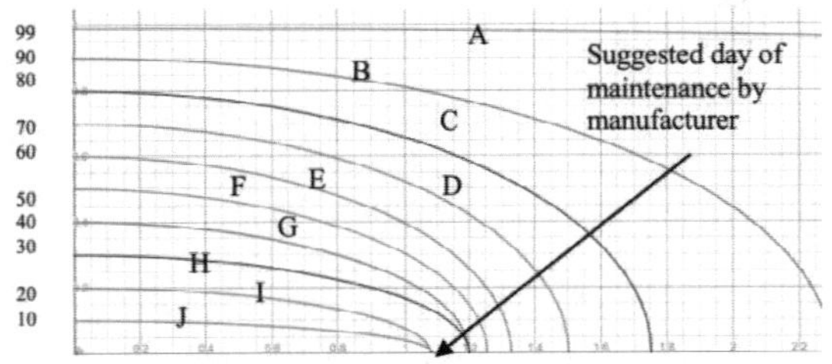

Figure 5: Reliability Condition (%) vs Years

From Figure 3 it can be seen that the CBM system suggests maintenance at the time when the manufacturer suggests, if the condition of the machinery was at 10-20% reliability when the data gathering activity started. This is very unlike to be true since less than half of the maintenance time interval suggested by the manufacturer has passed. That means that the time of maintenance could be increased further of what the manufacturer suggests. For example, if it is assumed that the machinery has 50% reliability when the data gathering started then the maintenance can wait up to almost two months. In the long term the time gained can accumulate to be many months, even years.

Figure 4 is the same plot as figure 3 but for DG2.

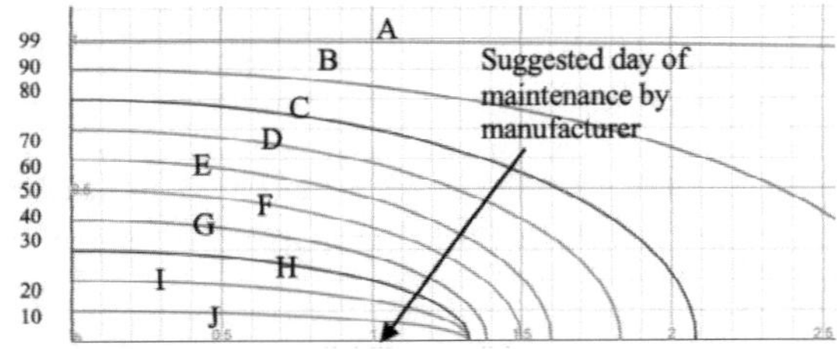

Figure 6: Reliability Condition (%) vs Years

From Figure 4 it can be seen that the CBM system suggests that failure will be later than when the manufacturer suggests for any initial reliability assumption. If it is assumed that the machinery has 50% reliability when the data gathering activity started, the gain in time of maintenance will be six months, which is very significant.

Table 4 shows for DG1 and DG2 the time difference between the model's prediction and manufacturer's prediction for different assumed conditions of the machinery counting from when the data gathering activity started.

Table 2: Time difference from suggested maintenance

Reliability	DG1	DG2
99.9%	9 yrs-2 m	11 yrs-5m
90%	1 yrs-3 m	1 yr-10m
80%	8 m	1 yr-1m
70%	5 m	10 m
60%	3 m	7 m
50%	2 m	6 m
40%	1 m	5 m
30%	1 m	4 m
20%	0 m	4 m
10%	0 m	4 m

From Table 4 it can be see that for DG1 cylinders the gain in time of performing maintenance can be very significant as the reliability of the machinery is assumed higher when the data gathering activity started. The manufacturer's recommendation for maintenance are valid only if the condition of the machinery is as low as 10-20% when the data gathering started.

For the DG2 cylinders it is even more obvious that the gain in time is very significant since even the lowest assumption of reliability (10%) gives extra 4 months until maintenance. The gain in operational time as the assumed reliability is higher is very significant reaching even a year of extra time, (e.g. when the reliability is at 80%).

4.7 Discussion of Case Studies

From the results of the case studies in the maintenance problem of the DG1 and DG2 cylinders is clear that the proposed CBM system would give an increased operational time from what the manufacturer suggested. This shows that the estimations for the maintenance intervals from the manufacturers are very conservative and do not take into account the real condition of the machinery at each maintenance interval. Thus, the proposed CBM system could greatly benefit the shipowners and operators by giving them the ability to be in charge of the maintenance schedule and reduce significantly the amount of times that maintenance needs to be performed and increase the operational time of the machinery.

More specifically from the case studies:

In the maintenance problem of the DG1 cylinders the CBM system showed that the reliability of the machinery must be very low in order to justify maintenance according to the manufacturer.

In DG2 the CBM system identified that there was no reason to perform maintenance in the cylinders when the manufacturer suggests since even for the lowest assumed reliability the time of failure is months after when the manufacturer suggests.

Overall, both cases highlight the weakness of using fixed time intervals, which are usually too strict, for safety reasons since there is no real time assessment of the machinery and they need to account for every possible deterioration pattern.

5 Conclusions

Current literature has shown that the marine sector is behind in implementing CBM for maintenance of the machinery systems, mostly because of the non-existence of an advanced CBM with simple architecture and effectiveness. This paper introduced a new CBM framework consisting of an Object-Oriented Database, machine learning techniques and a diagnostic and prognostic system.

The results of the case studies showed that there is a big room of improvement from just using the maintenance intervals suggested from the manufacturers. The proposed CBM is taking into account the condition of the machinery in real time to provide the shipowners with valuable information that can save them not only valuable time but also improve the safety of everyone onboard.

6 Future Research

In future scientific publications the detailed design of the CBM system will be presented including, but not limited, to how sensor values will be taken and stored by the CMD, how the CMD understands different operational profiles in real time, how the algorithm will be trained on-line by using the CMD data and more details of the mathematics behind the innovative failure model that performs prognostics.

In addition, testing of the whole CBM framework will be done again with further data collected from sensors on-board vessels and results will also be presented in future publications.

7 Acknowledgments

This partnership received partial financial support from the Knowledge Transfer Partnerships (KTP) programme. KTP aims to help businesses to improve their competitiveness and productivity through the better use of knowledge, technology and skills that reside within the UK knowledge base. This successful Knowledge Transfer Partnership project, funded by UK Research and Innovation through Innovate UK, is part of the government's Industrial Strategy.

References

1. DNV GL, (2014). *Beyond Condition Monitoring in the Maritime Industry*. DNV GL Strategic Research & Innovation Position Paper 6-2014.
2. Raptodimos, Y & Lazakis, I. (2018) Using artificial neural network- self-organising map for data clustering of marine engine condition monitoring applications, Ships and Offshore Structures, 13:6, 649-656, DOI: 10.1080/17445302.2018.1443694
3. Bzdok, D. (2018). *Statistics versus Machine Learning*. Nature Methods-April 2018.
4. Nowlan, F.S. and Heap, H.F. (1978). *Reliability-centred maintenance*. United Airlines, San Francisco, 1978.
5. Broberg (1973). Broberg's report cited in Failure Diagnosis & Performance Monitoring, Vol. 11 edited by L.F. Pau. Marcel-Dekker, 1981.
6. MSDP Studies (1993). *Age Reliability Analysis Prototype Study*. American Management Systems, U.S. Naval Sea Systems Command Surface Warship Directorate, USA.
7. Allen, T.M, (2001). *US Navy Analysis of Submarine Maintenance Data and the Development of Age and Reliability Profiles*. Submarine Maintenance Engineering, Planning and Procurement (SUBMEPP).
8. Lazakis, I., Gkerekos, C., & Theotokatos, G. (2018). Investigating an SVM-driven, one-class approach to estimating ship systems condition. *Ships and Offshore Structures*.
9. Gkerekos, C., Lazakis, I., & Theotokatos, G. (2017). Ship machinery condition monitoring using performance data through supervised learning. In *Proceedings of the 2017 Smart Ship Technology Conference* (pp. 105-111). London: Royal Institution of Naval Architects.
10. Raptodimos, Y., & Lazakis, I. (2016). An artificial neural network approach for predicting the performance of ship machinery equipment. In Maritime Safety and Operations 2016 Conference Proceedings (pp. 95-101).
11. Geron, A. (2017). *Hands-On Machine Learning with Scikit-learn & TensorFlow*. O'Reily Media.
12. Shorten, D.C. (2012). *Marine Machinery Condition Monitoring: Why has the shipping industry been slow to adopt?*. Lloyds Register, London, 2012.
13. Jardine A.K.S, Lin D, Banjevic D. (2006). *A review on machinery diagnostics and prognostics implementing condition-based maintenance*. Mechanical Systems and Signal Processing 2006;20:1483–510.
14. Shin, J-H. & Jun, H-B., (2015). *On condition-based maintenance policy*. Journal of Computational Design and Engineering 2, 119-127.
15. Poongodai, A. & Bhuvaneswari S., R., (2013). *AI Technique in Diagnostics and Prognostics. International Journal of Computer Applications*. 2nd National Conference on Future Computing February 2013.
16. Sugeno, M. and T. Yasukawa (1993). *A fuzzy-logic- based approach to qualitative modeling*. IEEE Transactions on Fuzzy Systems 1(1), 7–31.
17. Schwabacher, M. & Goebel, k. (2005). *A Survey of Artificial Intelligence for Prognostics*. NASA Amess Research Center MS 269-3.
18. Manno, G. and Knutsen, K.E. (2014). *An Importance Measure approach to System level Condition Monitoring of Ship Machinery Systems*. Conference paper-June 2014.
19. Zope Object Database, (2017). *ZODB Book Documentation Release 1*. https://media.readthedocs.org/pdf/zodb/latest/zodb.pdf, last assessed 2019/03/15.
20. Atzeni, Paolo & Atzeni, Paolo, 1957- & Ceri, Stefano & Paraboschi, Stefano & Torlone, Riccardo (1999). *Database systems : concepts, languages & architectures*. The McGraw- Hill Companies, London ; New York
21. Elmasri,R.andS.Navathe,B.(2010).*FundamentalsofDatabaseSystems*.AddisonWes- ley, 6th edition.
22. Silberschatz, A., Korth, F. H. and Sudarshan, S. (2011). *Database System Concepts*. Mc Graw Hill, 6th edition.
23. Ireland,C.andBowers,D.(2015).ExposingtheMyth:Object-RelationalImpedanceMis- match is a Wicked Problem.
24. Breunig M.M., Kriegel, H.P., Raymond T.Ng. and Sander, J. (2000). *LOF: Identifying Density-Based Local Outliers*. Proc. ACM SIGMOD 2000 Int. Conf. On Management of Data, Dalles, TX, 2000.
25. Lazarevic A., Kumar V. and Srivastava J. (2005). *Intrusion Detection: A Survey*. Chapter 2.
26. Campos G.O., Zimek A., Sander J. and Campello R.J. (2015). *On the Evaluation of unsupervised outlier detection: measures, datasets and an empirical study*. Data Mining and Knowledge Discovery.
27. Scikit-Learn, (2020). *Anomaly detection with Local Outlier Factor (LOF)*. https://scikit-learn.org/stable/modules/outlier_detection.html, last accessed: 2019/03/15).
28. Song H., Jiang Z., Men A. and Yang B. (2017). *A Hybrid Semi-Supervised Anomaly De- tection Model for High-Dimensional Data*. Comput Intell Neurosci. 2017; 2017: 8501683.
29. ABS (2016). *Guidance notes on equipment condition monitoring techniques*.
30. MAN B&W (2004). DMS2100i Bridge Manoeuvring System *ME/ME-C Engines User Manual*.

INTELLIGENT TRANSPORT SYSTEMS IN SHIPPING - MAKING IT (AND KEEPING IT) SECURE

D.A Kerigan-Kyrou, Partnership for Peace Consortium of Defense Academies, Germany

SUMMARY

Intelligent Transport Systems (ITS) and Intelligent Infrastructure Systems (IIS) have been with us for many years. But all too often ITS and IIS is poorly planned and badly integrated. However, the devices used - now often called the' Internet of Things'; IoT - which enable the 'next generation' of ITS are growing in number exponentially. In the maritime environment Smart Ship Technology presents an opportunity to greatly improve ITS / IIS, thereby increasing efficiency, both financial and also environmental. This paper will introduce how Smart Ship Technology can better interact with the vast growth in IoT in the entire supply chain. It will then focus on a new approach to enabling this integration to be built and maintained in a secure way.

1. WHAT ARE INTELLIGENT TRANSPORT SYSTEMS?

Intelligent Transport Systems' or 'ITS' is a system in which information and communication technologies - including infrastructure, vehicles and users, traffic management, mobility management, and interfaces between different modes of transport - are utilized. The idea of linking transport to other systems - a 'Systems of Systems' approach - such as logistics, environment, health, emergency services - was examined by Prof David King, the UK Government's Chief Scientific Advisor in the Foresight Directorate in 2006. This was described as Intelligent Infrastructure Systems - IIS. [1]

Unfortunately for the maritime industry the focus by governmental bodies didn't progress IIS in any regulated way. Indeed, what regulation has occurred has not only focused on ITS but this ITS itself has very much been road vehicle based, as encapsulated in Directive 2010/40/EU. [2] This is rather strange, because it is the maritime sector that that's been at the forefront in the development of technology. Indeed, sextants, nautical charts, lighthouses, beacons, buoys, tide gauges etc have been developed over hundreds of years,

And this lack of regulatory oversight has not prevented the development of ITS aboard ships, ashore within ports and within the wider logistical framework - indeed, the type of IIS predicted by Prof King in 2006. Maritime development has arguably led the way in ITS and what was known, back then, as IIS.

Since Prof King's report in 2006 the much more messy term the 'Internet of Things' - IoT - has gradually replaced the idea of Intelligent Infrastructure Systems. That is not good for security or long-term planning. The very term IoT implies that it's these bits and pieces - 'the things' which are now driving strategy, as opposed to strategy driving the implementation of 'the things' / devices. In others words, IoT is driving the system, not the other way round. But it is the IoT which is the central and integral part of all Intelligent Transport Systems.

2. THE INTERNET OF THINGS AND THE MARITIME ENVIRONMENT

IoT essentially consists of internet devices receiving and transmitting data. These devices contain sensors and actuators able to perform critical functions. Most of these devices are, in effect, computers running software and 'firmware' (a computer program stored within the hardware).

Modern vessel components increasingly comprise the Internet of Things connected to control systems via the internet. The list of IoT aboard and ashore is growing exponentially. For example, the power management, loading and stability, container monitoring, alarms, bridge control console, Electronic Chart Display and Information System (ECDIS), Automatic Identification System, Navigation Decision Support (NAVDEC), Voyage Data recorders, Computerized Automatic Steering, and the Global Maritime Distress and Safety System (GMDSS). Ports increasingly feature multiple examples of IoT including port security, access control, CCTV, gates, ID cards, automated cargo handling equipment, the Terminal Operating Centre, cranes, and integrated supply chain logistical systems. (Modern ports are rapidly becoming advanced logistical centres incorporating IoT into almost every function).[3]

Moreover, port IoT devices are directly interacting with vessels' IoT including communications, GPS (Global Positioning System), lock operations, maintenance and management, pollution and environmental control systems.[4,5]

In the maritime environment one of the biggest emerging challenges is the security of the 'Internet of Things' (IoT), operated by Industrial Control Systems.[6]

The US and EU have highlighted particular concerns regarding the 'security vulnerability' of these devices[7].

A proposed US Bill ,'The Internet of Things Cybersecurity Improvement Act of 2017', defines a security vulnerability as a "compromise of the confidentiality, integrity, or availability" of a device or its information.[8]

These IoT concerns clearly extend to the maritime environment. April Danos of Port Fourchon Louisiana and the US National Maritime Security Advisory Committee, and a leading authority on maritime cybersecurity, underscores the challenge: "We are blind, useless and potentially locked out of our own house if we are hacked; and let's face it, it isn't 'if', it's 'when'."[9]

To be blunt: A very significant proportion of maritime IoT is insecure by design. And the maritime industry - at every level - has comprehensively failed to understand the immense cyber risk to shipping. This is as true with port operators as it is with vessel operators. The primary reason for this is that cybersecurity is compartmentalised as an 'IT' function. These IT managers across the maritime environment have, with very few exceptions, failed to realise that cybersecurity is an issue which is of concern to everyone who works in the maritime sector. But getting this basic message across has been near impossible, especially to the boards in charge of shipping and port operators.

The vast majority of cybersecurity breaches are not seen by the victim as the nefarious actor does not wish to be identified and indeed have their breach spotted. A fundamental misunderstanding of cybersecurity is that breaches can be identified. Most of the time they remain undetected, normally for years, if they are ever discovered at all.

Indeed, Stefan Lüders of CERN, one of the world's lead authorities on the security of IoT states that while hundreds of their IoT devices control power, security and research, 32% of CERN's IoT devices either crashed or failed when faced with the most basic vulnerability scan.

There have already been allegations of the 'spoofing' of both military and civilian shipping.[10]

But this is the tip of the iceberg. To paraphrase US Col Pat Duggan (who was speaking about the military environment), what do we do when the entire maritime environment becomes a wireless node, constantly emitting or receiving data from innumerable locations across oceans for limitless reasons?[11]

3. SOLUTION: FIX THE SUPPLY CHAIN. CHANGE THE APPROACH.

So what's the solution? Firstly, it is understood that cybersecurity is a matter which concerns everyone.

US RADM Paul F. Thomas argues that managers and directors need to understand that cybersecurity is as integral to successful operations as logistics, fuel supply, or safety. Indeed, incorporating cybersecurity into safety and security, rather than having it perceived as an isolated cost centre, is central to the development of maritime cybersecurity.[12]

And as Col Duggan states: "We must change our institutional biases toward cyber. For too often when it comes to cyber discussions we get over focused on the technical, and leaders get uncomfortable...I think we have misplaced 'cyber' as simply an intell or a signals thing, when the fact of the matter is that cyber is an everybody thing, and every single military branch, service or discipline has a critical role to play." Exactly the same can be applied to the civilian environment and the entire supply chain of IoT.[13]

The UK National Cyber Security Centre has specifically stated that the supply chain and 'Managed Service Providers' are a key way for a nefarious actor to breach a target. Therefore we need to be ensuring that the entire supply chain in naval architecture - whether it is the manufacture of devices or the provision of services which enable the construction of maritime vessels - becomes secure. There is some legislation to help this, but on its own it is insufficient. The EU Directive on Network and Information Systems, 2016,[14] is a step in the right direction. Moreover, the IMO cybersecurity recommendations: IMO Guidelines on Maritime Cyber Risk Management (MSC-FAL.1/ Circ.3),[15] although non-binding are an excellent step in the right direction in securing the supply chain as the Guidelines "encourage administrations to ensure that cyber risks are appropriately addressed in existing safety management systems... no later than the first annual verification of the company's Document of Compliance after 1 January 2021." However as Parker and Kerigan-Kyrou argue, January 1, 2021 is not a deadline in any meaningful sense, unless regional administrations make its recommendations mandatory. [16] So there needs to be mandatory enforcement of supply chain security at the EU Maritime Safety Agency and the US United States Maritime Administration in order to create and spur a global approach. The IMO is unlikely to be able to do this on its own without strong support from the US and the EU.

4. CONCLUSIONS

The Maritime industry has been at the forefront of development ITS. But regulation has not kept up keeping these systems secure, whether at sea, ashore or in the supply chain. An integrated approach between the maritime industry and regulators in the US, EU, and across the world is needed soon.

5. REFERENCES

1. Intelligent Infrastructure Futures Project Overview. 2006. Available at: https://assets.publishing.service.gov.uk/government/uploads/system/uploads/attachment_data/file

/300334/06-522-intelligent-infrastructure-overview.pdf).

2. Directive 2010/40/EU, at: https://eur-lex.europa.eu/legal-content/EN/ALL/?uri=CELEX%3A32010L0040
3. NATO Maritime Interdiction Operational Training Centre Journal. 'Applying the NATO Cybersecurity Generic Reference Curriculum to the Maritime Environment' 2017. Kerigan-Kyrou D.A, Irish Defence Forces: An Cosantóir, 'Defining Our Approach to Cybersecurity', 2017.
4. Kerigan-Kyrou, NMIOTC 2017, op cit.
5. Security Industry Association Presentation, March 16, 2016 ‘Keeping Cargo Moving: Maritime Cybersecurity’ with Brett Rouzer, US Coast Guard Cyber Command, and April Danos, Port Fourchon, Louisiana. Available at: https://aapa.files.cms-plus.com/AAPAPresentations/AAPA%20PSGP%20Webinar%20-%20Cybersecurity%20-%20April%20Danos.pdf
6. For further explanation of SCADA (Supervisory Control and Data Acquisition), and Industrial Control Systems see: Robert Radvanovsky & Jacob Brodsky, ed., ‘Handbook of SCADA / Control Systems Security’ (Boca Raton: CRC Press), 2016
7. EU ENISA: ‘Security and Liability in the Internet of Things’, June 2017.
8. United States Senate. 115th Congress, 1st Session. ‘Internet of Things Cybersecurity Improvement Act of 2017’. See also concerns raised about IoT security by Senators Mark R. Warner and Cory Gardner, co-chairs of the Senate Cybersecurity Caucus, at: www.warner.senate.gov/public/index.cfm/pressreleases?id=06A5E941-FBC3-4A63-B9B4-523E18DADB36
9. ‘Keeping Cargo Moving: Maritime Cybersecurity’ op.cit.
10. UT Austin Researchers Successfully Spoof an $80 million Yacht at Sea,” UTNews, 29 July 2013 at https://news.utexas.edu/2013/07/29/ut-austin-researchers-successfully-spoof-an-80-million-yacht-at-sea and: “2017-005A-GPS Interference-Black Sea MARAD Alert,” United States Department of Transportation at https://www.marad.dot.gov/msci/alert/2017/2017-005a-gps-interference-black-sea/
11. Original quote from Col Pat Duggan can be found in: Kerigan-Kyrou D.A. Irish Defence Forces - An Cosantóir. Cybersecurity & the Internet of Things - Transforming Our Approach to Defence.
12. RADM Paul F. Thomas, US Coast Guard. Evidence to: ‘Protecting Maritime Facilities in the 21st Century’. Hearing before the Subcommittee on Border and Maritime Security; US House of Representatives, October 8, 2015. Serial No. 114–35, p.43.
13. Georgetown University International Conference on Cyber Engagement, April 24, 2017.
14. www.ec.europa.eu/digital-single-market/en/network-and-information-security-nis-directive), and IMO: http://www.imo.org/en/OurWork/Security/Guide_to_Maritime_Security/Pages/Cyber-security.aspx
15. IMO Guidelines on Maritime Cyber Risk Management (MSC-FAL.1/Circ.3). July 5, 2017. Available at: http://www.imo.org/en/OurWork/Facilitation/docs/FAL%20related%20nonmandatory%20instruments/MSC-FAL.1-Circ.3.pdf
16. See: C.J. Parker and D.A Kerigan-Kyrou: NATO Maritime Interdiction Operational Training Centre Journal. 'A Decade of Disruption: Cyber in the Maritime Domain',' 2020.

6. AUTHORS BIOGRAPHY

Dinos A. Kerigan-Kyrou is an instructor on the NATO Defence Education Enhancement Programme, based at the Partnership for Peace Consortium (PfCP), Garmisch-Partenkirchen. He is responsible for the cybersecurity training on the Joint Command & Staff Course at the Irish Defence Forces, and is one of the co-authors of the NATO / PfPC Cyber Security Training Curriculum. He was recently made the Visiting Research Fellow on the Ethical Hacking & Cyber Security Degree at Abertay University, Scotland. He is a Chartered Member of the Institute of Logistics & Transport.

MACHINE LEARNING AND CLOUD COMPUTING FOR REMOTE MONITORING OF WAVE PIERCING CATAMARANS: A CASE STUDY USING MATLAB ON AMAZON WEB SERVICES

B. Shabani, **J. Ali-Lavroff**, **D. S. Holloway**, University of Tasmania, Australia

S. Penev, UNSW Sydney, Australia

D. Dessi, Italian National Research Council, Italy

G. Thomas, University College London, UK

SUMMARY

Wave load cycles, wet-deck slamming events, accelerations and motion comfort are important considerations for high-speed catamarans operating in moderate to large waves. This paper provides an overview of data analytics methods and cloud computing resources for remotely monitoring motions and structural responses of a 111 m high-speed catamaran. To satisfy the data processing requirements, MATLAB Reference Architectures on Amazon Web Services (AWS) were used. Such combination enabled fast parallel computing and advanced feature engineering in a time-efficient manner. A MATLAB Production Server on AWS has been set up for near real-time analytics and execution of functions developed according to the class guidelines. A case study using Long Short-Term Memory (LSTM) networks for ship speed and Motion Sickness Incidence (MSI) is provided and discussed. Such data architecture provides a flexible and scalable solution, leading to deeper insights through big data processing and machine learning, which supports hull monitoring functions as a service.

1. INTRODUCTION

This paper provides an overview of data analysis methods and cloud computing resources being considered for Volcan de Tagoro (see Figure 1), which has operated in Spain, Canary Islands in the route between Las Palmas and Tenerife since August 2019. Classified by DNV GL, the vessel has a maximum deadweight (DWT) of 1000 tonnes, can carry up to 1200 crew and passengers and 401 cars, and has a specification speed of 42.4 knots at 600 tonnes DWT.

Large wave piercing catamarans (WPCs) have the capability of satisfying economic demands for fast sea transportation with improved seakeeping characteristics compared to conventional catamarans [1, 2]. Since the 1990s many large WPCs have been constructed at Incat Tasmania's Hobart shipyard, ranging from 74 m to 112 m in overall length and being operated mainly as passenger/vehicle ferries. These aluminium vessels have a distinctive design due to an above-water centre bow connecting two demihulls at the forward region. The propulsion systems of the Incat Tasmania WPCs are water jets and an active ride control system is installed to improve passenger comfort. Figure 1 shows the 111 m wave-piercing catamaran (Volcan de Tagoro, Hull 091) built by Incat Tasmania in 2019. Table 1 shows the main particulars of Hull 091. A hull monitoring system was developed for Hull 091 as previously described in [3] to continuously monitor the ship motions and structural responses.

It should be noted that wave load cycles, wet-deck slamming events, accelerations and motion comfort are important considerations for high-speed catamarans when operating in moderate to large waves. A guideline [4] presents the standard components to receive a Hull Monitoring System Notation (HMON) for different vessel types including high speed light craft. Although developing a hull monitoring system according to the classification guidelines for such vessels is broadly acceptable, the data processing requirements can be as sophisticated to implement as the system components themselves. One needs to consider rainflow counting, filtering, probability distribution, fatigue damage estimation, warning due to slamming and many others. Machine Learning (ML) and Deep Learning (DL) algorithms [5-8] add a layer of intelligence to onboard monitoring systems, with their potential yet to be fully explored. However, ML and DL data pipelines will also create more complexities for such systems, as is also evident from the class recommended practices for data quality assessment (i.e. DNVGL-RP-0497) and data-driven algorithms and models (i.e. DNVGL-RP-0510).

One solution to overcome complexities in big data processing and machine learning of modern hull monitoring systems is cloud computing. This represents an alternative approach to onboard processing and storage. Both methods provide many advantages, with the latter being a common method within the industry. However, cloud computing resources are an excellent choice for big data processing [9] and real-time analytics and their applications across many industries are well understood. A cloud processing solution considering MATLAB Reference Architectures on Amazon Web Services (AWS) is introduced in this paper. Such a combination facilitated parallel processing of raw data and provided a framework for training various ML and DL algorithms on the cloud platform.

Figure 1 Volcan de Tagoro, Hull 091 111m wave-piercing catamaran built by Incat Tasmania in 2019.

2. HULL MONITORING SYSTEM

A hull monitoring system has been developed for Hull 091 Incat vessel (see Figure 1). The vessel was instrumented in July 2019 using a motion reference unit (MRU), a bow accelerometer and 10 strain gauges. An ultrasonic wave sensor was placed in the bow area to measure the incident wave profiles. An overview of the remote monitoring system is shown in Table 2. Figure 2 shows the approximate longitudinal locations of the cDAQ, laptop, MRU, strain gauges, bow accelerometer and the ultrasonic sensor. The sampling rates were set to 100 Hz for the bow and MRU acceleration data, 50 Hz for MRU heave, pitch, and roll data and 9 Hz for the ultrasonic sensor. The National Instruments (NI) module sampling rate was also 1000 Hz for strain measurements (due to a minimum requirement of NI-9236 module) but it was down sampled to 100 Hz before recording data to files.

Table 1 Main particulars of Hull 091

Length overall	111.9 m
Length waterline	103.2 m
Beam Overall	30.5
Draft	4.1 m
Demihull beam	5.8 m
Max Deadweight	1000 tonnes
Speed	42.4 knots

Table 2 Overview of Hull 091 remote monitoring components

Category	Item	Quantity	Description
Sensors /data source	Strain gauges	10	HBM 1-LY43-6/350
	Accelerometer (3-axial)	1	CrossBow- CXL04GP3
	Ultrasonic sensor	1	ToughSonic 50(TSPC-21SRM-485)
	Motion reference unit (MRU)	1	SBG Systems- Ellipse2-A
	Global Positioning System (GPS) receiver	1	Hull 091 GPS distributor
Data acquisition	Data acquisition (DAQ) module	1	National Instrument (NI cDAQ-9174)
	Strain gauges module (8 channel)	2	National Instrument (NI-9236)
	Universal Input module (4 channel)	1	National Instrument (NI-9219)
Computer, accessories and software	Laptop	1	Dell latitude 7490
	USB Hub	1	Powered USB hub
	Onboard Monitoring Software	1	Customised LabVIEW program
	Remote Desktop Access	1	TeamViewer
	Laptop tray	1	RAM Universal tray (RAM-234-3)
	Cabinet	1	PCLocs – Carrier 10
	NMEA to USB convertor	1	Digital Yacht
Storage & connectivity	External hard drive	1	Samsung- USB-C 1T SSD
	Cloud based storage	2 TB (scalable)	Google Drive/AWS S3
	WiFi/LAN router	1	Digital Yacht 4G Connect PRO
	4G antennas	2	Digital Yacht 4G Connect PRO
	Sim Card	1	Simyo 4G

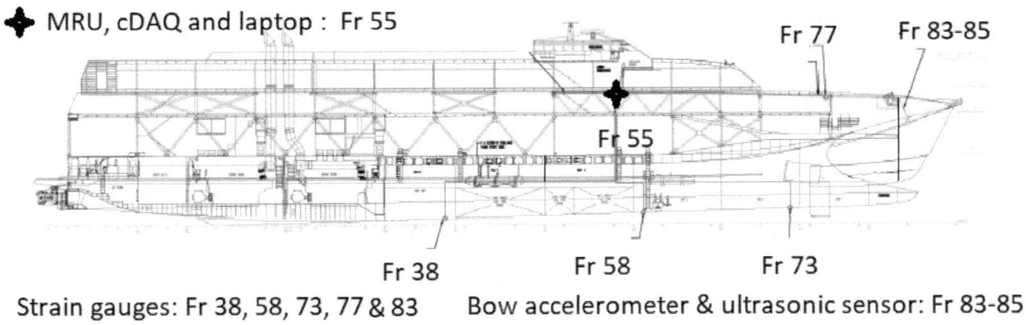

Figure 2 Hull 091 cabling diagram from frame 55 to frames (Fr) 38, 58, 73, 77, 83 and 85, showing the approximate longitudinal locations of the National Instruments cDAQ, laptop computer, MRU, strain gauges, bow accelerometer and ultrasonic sensor listed in Table 2.

A LabVIEW program was developed to record and upload data automatically to a cloud storage. It is worth mentioning that developing a customised LabVIEW program enabled the integration of the MRU and GPS data. This was achieved by incorporating SBG Systems (Ellipse2-A) and NMEA-0183 GPS drivers, to the strain gauges, bow accelerometer and ultrasonic sensor data acquired from NI 9236 and 9219 modules.

Furthermore, a file management function was included so that the files could be uploaded into the Google Drive and then deleted from the computer hard drive. The choice of Google Drive here is due to it having a general purpose, temporary storage for preliminary data access and some other user-friendly features such as automatic synchronisation and backup to manage large data files. The customised upload code developed in the LabVIEW program can be adjusted to upload the data into a cloud platform which enables advanced features such as machine learning, deep learning and near real-time monitoring. Such features, for example, are available in public cloud computing services such as Amazon Web Services (AWS) Google Cloud Platform (GCP), and Microsoft Azure.

3. CLOUD COMPUTING AND PARALLEL PROCESSING

3.1 MATLAB REFERENCE ARCHITECTURES ON AMAZON WEB SERVICES (AWS)

As a public cloud service provider, among many others, AWS provides a broad range of services including computation, storage, analytics, machine learning, Internet of Things, and security. The full list of all services is available from the AWS management console. One of the services that provides all required resources in a specific, predefined format is AWS CloudFormation.

Figure 3 shows two different solution architectures using MathWorks products on AWS. The CloudFormation templates for each specific solution are available from the MathWorks page on GitHub [10]. Figure 3a presents a simple architecture in which the user can set up MATLAB Desktop on Amazon Elastic Compute (EC2) for a scalable cloud computing. The connection to the compute instance which is in a Virtual Private Cloud (VPC) is available by a Remote Desktop Protocol (RDP) client software and Secure Shell (SSH) tunneling, with the username and password to be specified during the launch of the template.

The access of the compute instance to other resources such as Amazon Simple Storage Service (S3) will need to be enabled by attaching specific policy to the compute instance. Various types of EC2 instances are available to select, including general purpose such as M5 series, compute optimised such as C5 series, and accelerated computing such as P3 series. The accelerated computing type comes with graphics processing units (GPU), which is ideal for training deep neural networks and allows to significantly reduce training times when compared with Central Processing Units (CPU). It should be stressed that the MathWorks reference architectures are pre-configured to use single or multiple GPUs and this facilitates the process on both AWS and Microsoft Azure cloud services.

(a)

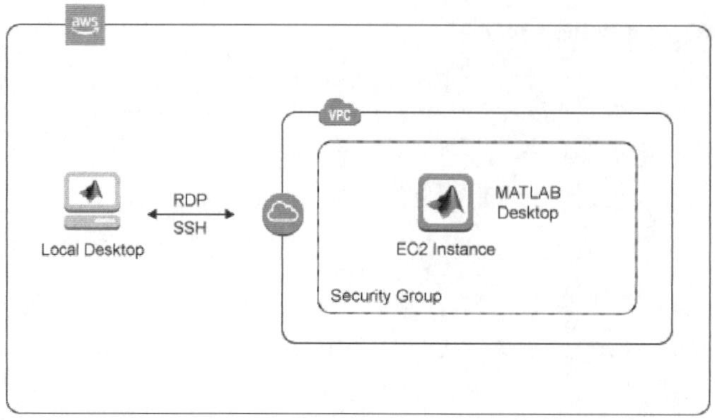

(b)

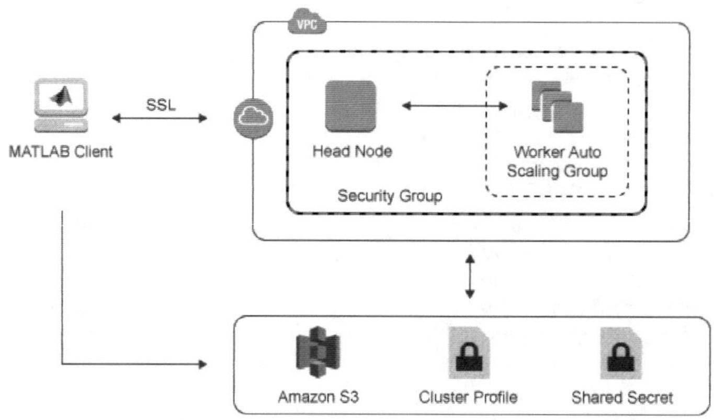

Figure 3 MathWorks reference architectures on AWS (top) MATLAB on AWS (bottom) MATLAB Parallel Server on AWS

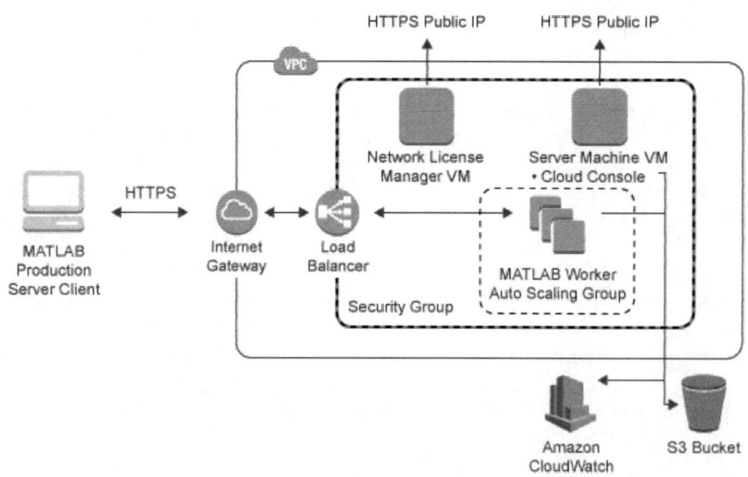

Figure 4 MATLAB Production Server on AWS

Parallel CPU processing is often required when training various machine learning algorithms and this can be performed by MATLAB Classification Learner or Regression Learner Apps. MATLAB recommends a minimum of 2 virtual Central Processing Units (vCPU) for each worker because two vCPUs share only one Floating Point Unit. To give an example, by selecting m5.xlarge, which comes with 4 vCPUs and 16 GB memory, only 2 workers are available in the cluster, while by selecting m5.24xlarge as a compute instance on AWS, which comes with 96 vCPUs and 384 GB memory, a maximum of 48 MATLAB workers are available to be used in the local parallel processing.

The second architecture (Figure 3b) presents another approach for parallel processing and scaling of the compute workers. In this method, MATLAB workers are distributed across two or several EC2 instances with a dedicated EC2 instance as the head node. The number of workers is adjustable by changing the desired capacity within the autoscaling policy and the connection to the head node is enabled via SSL. This can present a convenient model when there are multiple users and these users have MATLAB on their local computers as they can connect to the parallel processing cluster remotely and perform their analytics. When dealing with files stored in an S3 bucket, access to the bucket should be enabled for each worker by attaching an S3 access policy to the autoscaling group policy.

Both architectures presented in Figure 3 are supported with MATLAB online licensing, which means individual account holders can setup their own resources on AWS. These architectures were used for processing raw data received from Hull 091 and training ML/DL algorithms to be briefly introduced in Section 4.

3.2 PRODUCTION SERVER

MATLAB Production Server on AWS was considered for real-time analytics and automated data processing. The reference architecture as shown in Figure 4 requires additional cloud resources in comparison with the two earlier architectures discussed. It can for example provide (i) a framework to automate data processing in relation to some customised monitoring functions such as MSI, rain flow cycle counts and peak over threshold; (ii) a ML production pipeline by hosting ML and DL classifiers; and (iii) a mechanism for calling MATLAB functions deployed on the production sever from either a MATLAB client or third-party clients/applications developed in other programming languages for realtime analytics or visualisation. More details of this architecture are given in [10].

Moreover, the analysis process can be fully automated by incorporating serverless resources such as AWS Lambda and DynamoDB for event triggering, concurrent executions and storing processed data. Such data architecture can provide a flexible, scalable, and cost-effective solution, leading to deeper insights through automated data processing. Consequently, it has been proposed that the hull monitoring functions and classifiers can be offered as a service, supporting Hull 091 and future vessels.

It should be noted that this study did not consider several options, and optimal solution is often not readily available. Therefore, it is encouraged to explore various options including software provided by the open source community and other packages. Amazon SageMaker Studio is also recommended for training and deploying ML algorithms as it simplifies compute resource allocations, experimenting with various models and hyper parameter tuning for optimal performance.

4. DATA ANALYSES

4.1 OVERVIEW

Developing appropriate functions and methods for analysing data collected from Hull 091 on the cloud is an ongoing project. The purpose of this section is to provide an overview of some standard analyses according to DNV GL guidelines and to highlight some selected analysis methods published in the literature for high speed vessels. This is important as the DNV GL guidelines do not specify signal processing requirements such as filtering, slamming, and whipping identification for such vessels.

For the current project three main areas have been considered for the analyses: (i) evaluating motion comfort through MSI (ii) identifying rainflow cycles for fatigue damage analyses, and (iii) developing a method for slamming identification

4.2 MOTIONS AND MOTION COMFORT

A function in MATLAB was developed to obtain weighted accelerations, according to ISO standard 2631-1. The outputs were then used to calculate MSI according to the vertical root mean square weighted acceleration (a_w) and exposure time (*T*) in seconds:

$$\mathrm{MSI} = \frac{1}{3} a_w \sqrt{T} \quad (1)$$

The above equation is a measure of whole-body exposure to low frequency vertical accelerations. Another parameter can be considered for higher frequency accelerations. This is the Vibration Dose Value (VDV) as recommended by DNV GL, which includes accelerations in x-, y- and z-directions. It should be noted that MSIs at other locations can be calculated by considering rigid body motions. Moreover, the vertical and lateral accelerations, pitch and roll can also be considered in evaluating passenger comfort and safety according to some threshold values given in [11]. The method presented in [12] to evaluate motion response amplitudes has also been considered. This requires the Fast Fourier Transform (FFT) of heave, pitch, roll and wave response amplitudes. For Hull 091, a helper function needs to be developed to estimate the wave elevation by relative displacement at the location of the bow ultrasonic sensor. More details for the estimation of wave spectra using an ultrasonic altimeter can be found in [13].

4.3 STRESS CYCLES AND FATIGUE ANALYSES

The DNV GL procedure for fatigue assessment of ship structures (DNVGL-CG-0129) has been considered for the analyses. The objectives of the analyses are to estimate fatigue rate, defined as the ratio of the measured fatigue damage and the budget damage per unit time. The stress cycle distribution was calculated using the rain flow cycle counting method according to ASTM Standard E-1049 by incorporating the MATLAB rainflow function. The outputs include amplitude and mean stress levels for each cycle, identified through the procedure. More discussions on fatigue and stress distributions for high speed catamarans are given in [14-16].

4.4 SLAMMING

Three approaches are mentioned in the rules for slamming identification, using (i) pressure transducers, (ii) an accelerometer in the bow area, or (iii) whipping responses from the global strain sensors. The main objective is identifying slam-induced bending stresses acting on either local or global structures. Providing an early warning due to slamming in terms of probability of slamming, reporting the number of slam peaks crossing certain thresholds and determining the location of slamming are desirable analyses.

Methods of analysing slam pressures and loads in irregular waves can also be found in [17, 18]. Wet-deck slamming loads pressures, slam induced bending moments, and slamming kinematics of wave piercing catamarans were investigated in regular waves [19-24]. Slamming identification methods for high speed crafts are also summarised in [25], from which the stress derivative threshold [26, 27], whipping-based [28] and fatigue-based criteria [25, 29] can be highlighted. A ML approach for classifying slamming events is also proposed in [3].

4.5 STATISTICAL ANALYSES AND EXTREME DESIGN LOADS

According to DNV GL, configuring threshold values, alarm levels, statistical calculations, probability distribution analyses, and trend predictions are other requirements of a standard hull monitoring system. For trend predictions, establishing a sequence of all features calculated from each individual sensor is required. Other analyses include obtaining several features such as descriptive statistics for each signal, mean zero crossing periods, peak to peak analyses, histogram of peaks, and distribution analyses. In addition, extreme value analyses and the probability of exceeding certain thresholds based on estimated probability curves need to be calculated. A method for predicting structural responses based on hull monitoring data is given [30]. Such analyses should also consider the combined effects of slam-induced and wave induced bending moments [31, 32] so that data driven models can be supported by Finite Element stress analyses and design load determination methods proposed in the literature.

5. LONG SHORT-TERM MEMORY NETWORKS

As mentioned earlier, establishing a sequence of features calculated from each individual sensor is required for trend predictions. Recurrent Neural Networks (RNNs) are for sequencing data and there have been numerous successful applications in various domains such as automated translation, natural language processing, speech recognition and signal processing [33, 34]. The proposal is to train Long Short-Term Memory (LSTM) networks, a variant of RNNs, by feeding data collected from Hull 091, and use MATLAB Production Server to host the LSTM classifiers for near real-time predictions. The network architecture consists of (i) an input layer (i.e. a sequence of data), (ii) an LSTM layer, (iii) a fully connected neural network layer and, (iv) an output layer (i.e. a regression layer).

Figure 5 shows an overview of an LSTM layer with LSTM blocks. Such a layer represents *C* number of features or channels, each with *S* length, where *S* represents the number of time steps. The LSTM layer has *S* LSTM blocks, with the first block requiring the feature vector at the first-time step and the initial state of the network to output *D* number of hidden units. Figure 5 also shows four components of each LSTM block: input gate (*i*), forget gate (*f*), cell candidate (g) and output gate (*o*). The mathematical operation of each component is given in Table 3. There are two activation functions: gate activate function, denoted by σ_g, and state activation function, denoted by σ_c.

Each component has its own input weights (*W*), recurrent weights (*R*) and bias (*b*) and they are referred to as learnable parameters of an LSTM layer. The cell and hidden states at time step *t* are,

$$c_t = f_t \odot c_{t-1} + i_t \odot g_t \quad (6)$$
$$h_t = \sigma_t \odot \sigma_c(c_t), \quad (7)$$

where $\odot$ denotes element-wise multiplication of vectors, c_t and h_t denote cell state and hidden state at time t and other variables are defined in Table 3. The default activations functions in MATLAB LSTM layer are a hyperbolic tangent function (tanh) and sigmoid function for σ_g and σ_c, respectively. Hyper parameters are learning rates, L_2 regularization factors for input weights, recurrent weights and biases and different initializers can be set for each W, R, and b matrices. The number of hidden units needs to also be specified by the user.

Table 3 LSTM block components and formula

Component	Formula	
Input gate	$i_t = \sigma_g(W_i\mathbf{x}_t + R_i\mathbf{h}_{t-1} + b_i)$	(2)
Forget gate	$f_t = \sigma_g(W_f\mathbf{x}_t + R_f\mathbf{h}_{t-1} + b_f)$	(3)
Cell candidate	$g_t = \sigma_c(W_g\mathbf{x}_t + R_g\mathbf{h}_{t-1} + b_g)$	(4)
Output gate	$o_t = \sigma_g(W_o\mathbf{x}_t + R_o\mathbf{h}_{t-1} + b_o)$	(5)

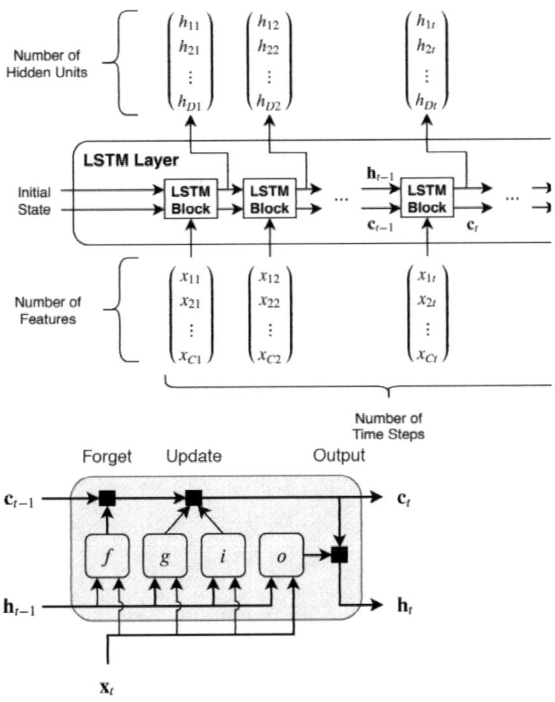

Figure 5 (top) LSTM Layer architecture (bottom) LSTM block architecture [35]

6. CASE STUDY

6.1 TRAINING AND TESTING

Short-term predictions of vessel speed and rms weighted acceleration are considered in this section as examples of using RNNs for the Hull 091 remote monitoring system. The aim is to train individual LSTM networks for each parameter, predict next step values and update the network state at each time step for improved predictions. An initial dataset consisting of all voyages in a month was selected and partitioned (i.e. the first 90% for training and the last 10% for testing the network). The window size for calculating the most frequent (i.e. modal) speed and rms weighted acceleration was 5 minutes, which is the length of each individual file received from the Hull 091 monitoring system. The rms weighted vertical acceleration was calculated using ISO standard 2631-1. Data preparation involved filling out missing data using 1-dimensional interpolation technique. Both parameters were standardized so that they have unit standard deviation and zero mean.

The individual networks were trained so that they can learn to predict the vessel speed and rms weighted acceleration at the next time step. Hyperparameters are learning rate, number of LSTM layers, number of hidden units in each layer, number of iterations or epochs, mini batch size, and threshold values for drop out layers. Over 20 initial combinations of network were selected, and the networks were trained on a single CPU to evaluate the performance of each, including the variations of training loss and training RMSE as a function of iterations during the training phase and also testing RMSE. A table hyperparameter was then created to perform hyperparameter tuning in a more systematic way by using MATLAB Experiment Manager on AWS by considering a P3 instance (i.e. a GPU accelerated computing resource).

A larger dataset was considered compared to the initial dataset by including voyages from September 2019 to June 2020. The train-test split ratios were 0.9-0.1 for this case study. Table 3 shows the range of hyperparameters considered. In the first experiment, a network with 200 hidden units, 3 LSTM layers, 0.002 learning rate and 200 epochs was the champion net among 24 trained networks. A second experiment with

12 options was then designed to select the final net as listed in Table 3. The networks were trained using the speed dataset. Hyperparameter tuning was not performed for rms weighted acceleration since it was seen that the final net was also a good choice for that. The training times on the single GPU varied from 53 seconds to 9 minutes depending on the training options considering about 24000 training datapoints. The choice for the optimiser was "Adam" [36] for all training experiments.

Table 3 Hyperparameter tuning details

	Hyperparameters	First experiment	Second experiment	Final
1	Learning rate	0.001, 0.002	0.002, 0.003	0.003
2	Number of hidden units	100, 200	200	200
3	Number of LSTM Layers	1,3,5	3	3
4	Max epoch	100, 200	200	200
5	Dropout Layer Threshold	0.25	0.25, 0.5	0.25
6	Minibatch Size	1	1,5,10	1
Total trained net		24	12	1

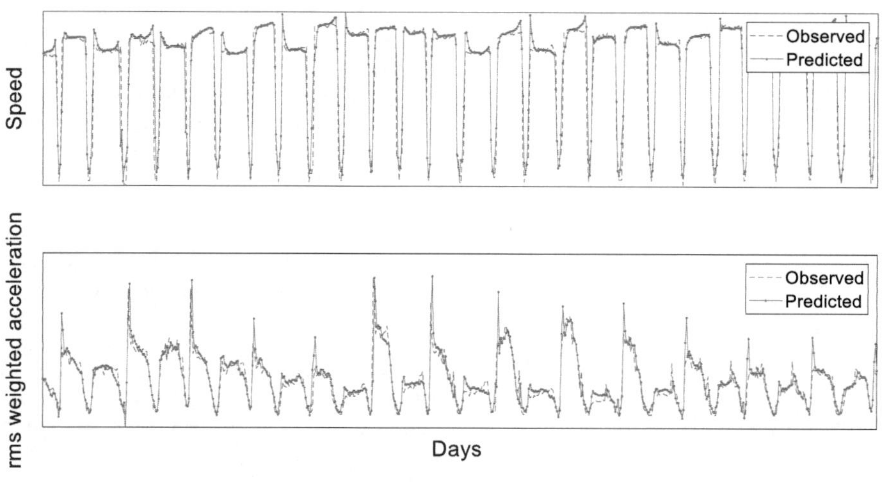

Figure 6 Observed speeds and rms weighted accelerations and predicted values for test voyages using trained LSTM networks

6.2 PERFORMANCE

Figure 6 compares predicted speeds and rms weighted accelerations with that observed. The predicted values in this figure are based on the observed values of the immediately previous time step, with a duration of 5 minutes between each time step. The predictions are generally consistent with those observed except for a few points in each voyage, particularly when the ship speed reaches the maximum level during the voyage. RMSE achieved for rms weighed accelerations and speeds were approximately 5% and 9% of the max values observed on the test dataset, respectively. It was found that RMSEs can be minimised by data clustering techniques, but the discussion on the approach is beyond the scope of the present study. Further investigations are recommended to benchmark the best possible performance of LSTM networks for various types of monitoring data as required by DNV GL guidelines in terms of trend predictions and sequencing.

6.3 DEPLOYMENT DISCUSSION

The deployment of DL trained regression models discussed earlier can be useful for two reasons relating to the influence of vertical accelerations on motion comfort. First, it can provide an early awareness of expected MSI and predicted speed so the ship captains can adjust the speed accordingly. They may decide to speed up, slow down or maintain speed based on the projected MSI and previous records. Second, a recommender system can be developed so that a suggested speed is returned according to a MSI threshold, which can either be a constant value or a dynamic value calculated for each voyage. The suggested speed needs further investigation to find an optimum.

In relation to the second point, a simplistic approach might be to suggest a lower speed (e.g. 5-10 %) than that projected if the expected MSI is above the threshold according to the industry standard threshold for passenger/vehicle ferries). If, for example, a 5% speed reduction is suggested when the predicted MSI is above the 95th percentile of all time MSI (i.e. for 5% of the time), this could affect the operational efficiency, so another consideration could be when to recommend a higher speed. On the other hand, how recommended speeds will improve rms weighed accelerations and the motion comfort due to the vertical accelerations should be evaluated. The exposure time, encounter wave frequency, relative vessel heading with respect to waves, arrival time, fuel efficiency and other operational factors are also important factors, and thus a full analysis needs to be conducted. Therefore, further research is required to find an optimal speed for passenger comfort by also considering ride control performance on MSIs at various speeds and headings. The optimal strategy is to be formulated as a solution to a multi-criterion optimisation problem.

Furthermore, it should be noted that the MSI should be calculated for various locations to better estimate the overall MSI. Ideally, the overall MSI should be as close as possible to the overall value obtained when the exact locations of seated passengers are known. This is because the interior design of the passenger deck can entice some passengers to be seated in specific locations or in the VIP section, and some may prefer to seat at extreme ends for better view of the sea.
The ISO equation for motion comfort due to whole body vibration is also questionable as it is a recommended metric for the general population. Frequent travellers may likely have better performance against possible sea sickness, in particular for a short exposure time (less than 2 hours).

The connectivity to the ship to the cloud cannot always be presumed and therefore the onboard monitoring system should be used as alternative when the connection to the production server is not available. However, cloud-based predictions and online reports are feasible at the beginning and at the final stages of each voyage. Such reports and dashboards can be valuable for day to day operations at sea in terms of motion comfort.

7. CONCLUSIONS

A remote monitoring system was proposed as an alternative to standard hull monitoring systems of high-speed catamarans and that can also apply to monohulls. This should be of interest to ship designers and builders for improved design load estimations and extreme values analysis of both global bending moments and slamming based on authentic full-scale in-service measured data in random seas. In addition, motion comfort and operational efficiency can be of importance to passenger/vehicle operators for day to day operations.
Several options are available for data engineering on a cloud computing environment. The data processing method proposed in this paper incorporated MATALB products on Amazon Web Services (AWS) including MATLAB Production Sever and Parallel Processing toolboxes, as well as other toolboxes for signal processing, machine learning and statistics.

A case study for trend predictions using Recurrent Neural Network (RNNs) was presented, in which sequences of speed and rms weighted acceleration were learned from measured data for the particular operational route of Hull 091. Due to the size of the dataset, GPU accelerated resources on AWS were used. Using MATLAB Experiment Manager, several Long Short-Term Memory (LSTM) network architectures were trained to obtain a final model.

Some data processing requirements and methods for slamming, fatigue and motion comfort were highlighted, mainly from DNV GL guidelines and published papers relevant to high-speed catamarans. Although MSIs can be used for motion comfort predictions as discussed, an optimal strategy is yet to be formulated as a solution to a multi-criterion optimisation problem, which includes vessel speed, rms vertical and lateral accelerations, encounter wave frequency, relative vessel heading with respect to waves, arrival time, fuel efficiency and other operational factors.

A complete data processing framework for high-speed vessels in random waves can be sophisticated. A systematic method of data processing that can include incorporating Machine Learning (ML) approaches is recommended, but it also requires investigations on feature engineering methods. A recommender system for improved motions and fuel efficiency is of interest for future investigation. Exploring novel techniques for onboard motion sickness evaluations, such as using deep neural networks and image processing, could also merit further considerations.

8. ACKNOWLEDGEMENTS

This work was undertaken in collaboration between the University of Tasmania, Revolution Design, Incat

Tasmania, University of New South Wales Sydney, Italian National Research Council and University College London through the support of the Australian Research Council Linkage Grant number LP170100555. The support of MathWorks team for providing licenses and consultation is appreciated.

9. REFERENCES

[1] Shabani B, Lavroff J, Holloway DS, Davis MR, Thomas GA. The influence of the centre bow and wet-deck geometry on motions of wave-piercing catamarans. Proceedings of the Institution of Mechanical Engineers, Part M: Journal of Engineering for the Maritime Environment. 2019;233:474-87.

[2] Fang C-C, Chan H-S. An investigation on the vertical motion sickness characteristics of a high-speed catamaran ferry. Ocean Eng. 2007;34:1909-17.

[3] Shabani B, Lavroff J, Holloway D, Penev S, Dessi D, Thomas G. Classifying bow entry events of wave piercing catamarans in random waves using unsupervised and supervised techniques. International Conference on Marine Industry 402019. p. 39-54.

[4] DNVGL. Rules for classification: Ships — Hull monitoring systems - HMON: DNVGL; 2018.

[5] Kubat M. An introduction to machine learning: Springer; 2017.

[6] Witten IH, Frank E, Hall MA, Pal CJ. Data Mining: Practical machine learning tools and techniques: Morgan Kaufmann; 2016.

[7] Wang J, Ma Y, Zhang L, Gao RX, Wu D. Deep learning for smart manufacturing: Methods and applications. J Manuf Syst. 2018;48:144-56.

[8] Ismail Fawaz H, Forestier G, Weber J, Idoumghar L, Muller P-A. Deep learning for time series classification: a review. Data Mining and Knowledge Discovery. 2019;33:917-63.

[9] Hashem IAT, Yaqoob I, Anuar NB, Mokhtar S, Gani A, Khan SU. The rise of "big data" on cloud computing: Review and open research issues. Information systems. 2015;47:98-115.

[10]GitHub. 2020. Mathworks Reference Architectures. [online] Available at: <https://github.com/mathworks-ref-arch> [Accessed 1 October 2020].

[11] Yun L, Bliault A, Rong HZ. High speed catamarans and multihulls: technology, performance, and applications: Springer; 2018.

[12] French BJ, Thomas GA, Davis MR. Slam occurrences and loads of a high-speed wave piercer catamaran in irregular seas. Proceedings of the Institution of Mechanical Engineers, Part M: Journal of Engineering for the Maritime Environment. 2015;229:45-57.

[13] Christensen KH, Röhrs J, Ward B, Fer I, Broström G, Saetra Ø, et al. Surface wave measurements using a ship-mounted ultrasonic altimeter. Methods in Oceanography. 2013;6:1-15.

[14] Nephi R. Johnson JPLMDC. Response and fatigue assessment of high speed aluminium hulls using short-term wireless hull monitoring. Structure and Infrastructure Engineering. 2017.

[15] Neuberg O, Drimer N. Fatigue limit state design of fast boats. Marine Structures. 2017;55:17-36.

[16] Teresa Magoga RO, Stuart Cannon, and Giles Thomas. Interdependencies between variables in fatigue analysis of a weightoptimised naval ship. First International Symposium on Risk and Safety of Complex Structures and Components: elsevier; 2019. p. 267-74.

[17] French BJ, Thomas GA, Davis MR. Slam occurrences and loads of a high-speed wave piercer catamaran in irregular seas. Proc Inst Mech Eng Part M J Eng Marit Environ. 2015;229:45-57.

[18] Davis MR, French BJ, Thomas GA. Wave slam on wave piercing catamarans in random head seas. Ocean Eng. 2017;135:84-97.

[19] Shabani B, Lavroff J, Davis MR, Holloway DS, Thomas GA. Slam loads and kinematics of wave-piercing catamarans during bow entry events in head seas. J Ship Res. 2018;62:134-55.

[20] Shabani B, Lavroff J, Holloway DS, Davis MR, Thomas GA. The effect of centre bow and wet-deck geometry on wet-deck slamming loads and vertical bending moments of wave-piercing catamarans. Ocean Eng. 2018;169:401-17.

[21] Shabani B, Lavroff J, Davis MR, Holloway DS, Thomas GA. Slam loads and pressures acting on high-speed wave-piercing catamarans in regular waves. Marine Structures. 2019;66:136-53.

[22] Shabani B, Lavroff J, Holloway DS, Davis MR, Thomas GA. Centre bow and wet-deck design for motion and load reductions in wave piercing catamarans at medium speed. Ships and Offshore Structures. 2019:1-17.

[23] Shabani B, Lavroff J, Holloway DS, Davis MR, Thomas GA. Wet-deck slamming loads and pressures acting on wave piercing catamarans. Int Shipbuild Prog. 2019;66:201-31.

[24] Lavroff J, Davis M, Holloway D, Thomas G, McVicar J. Wave impact loads on wave-piercing catamarans. Ocean Eng. 2017;131:263-71.

[25] Magoga T, Aksus S, Cannon S, Ojeda R, Thomas G. Identification of slam events experienced by a high-speed craft. Ocean Eng. 2017;140:309-21.

[26] Thomas GA, Davis M, Holloway D, Watson N, Roberts T. Slamming response of a large high-speed wave-piercer catamaran. Marine Technology. 2003;40:126-40.

[27] Jacobi G, Thomas G, Davis MR, Davidson G. An insight into the slamming behaviour of large high-speed catamarans through full-scale measurements. Journal of Marine Science and Technology. 2014;19:15-32.

[28] Dessi D, Ciappi E. Slamming clustering on fast ships: From impact dynamics to global response analysis. Ocean Eng. 2013;62:110-22.

[29] Thomas G, Davis MR, Holloway DS, Roberts T. The effect of slamming and whipping on the fatigue life of a high-speed catamaran. Australian Journal of Mechanical Engineering. 2006;3:165-74.

[30] Mondoro A, Soliman M, M.Frangopol D. Prediction of structural response of naval vessels based ona vailable structural health monitoring data. Ocean Eng. 2016:295-307.
[31] McVicar JJ, Lavroff J, Davis MR, Thomas G. Effect of slam force duration on the vibratory response of a lightweight high-speed wave-piercing catamaran. J Ship Res. 2015;59:69-84.
[32] Almallah I, Lavroff J, Holloway DS, Shabani B, Davis MR. Global load determination of high-speed wave-piercing catamarans using finite element method and linear least squares applied to sea trial strain measurements. Journal of Marine Science and Technology. 2019:1-13.
[33] Shrestha A, Mahmood A. Review of deep learning algorithms and architectures. IEEE Access. 2019;7:53040-65.
[34] Mu R, Zeng X. A review of deep learning research. KSII Trans Internet Inf Syst. 2019;13:1738-64.
[35] Mathworks.com. 2020. Long Short-Term Memory Networks- MATLAB & Simulink. [online] Available at:<https://www.mathworks.com/help/deeplearning/ug/long-short-term-memory-networks.html> [Accessed 1 October 2020].
[36] Kingma DP, Ba J. Adam: A method for stochastic optimization. arXiv preprint arXiv:14126980. 2014.

10. AUTHORS BIOGRAPHY

Dr Babak Shabani is postdoctoral researcher in the School of Engineering at the University of Tasmania, Australia. He is responsible for developing cloud-based machine learning systems specific to remote monitoring of high-speed catamarans. His previous experience includes "slam loads and kinematics of wave-piercing catamarans during bow entry events in head seas", the winner of Vice Admiral E. L. Cochrane Award (2019) from SNAME.

Dr Jason Ali-Lavroff currently holds the position of senior lecturer at the University of Tasmania. He is responsible for teaching in dynamics, vibrations, and design together with research on motions and loads of high-speed catamaran vessels. His previous experience includes extensive industry collaboration with Incat Tasmania on research involving ship motions and loads both at model scale as well as full-scale with a particular focus on slamming using both theoretical and experimental techniques. More recently this work has been extended by machine learning and classification at full-scale that is of particular relevance to the current paper.

Assoc Prof Damien Holloway holds the position of Associate Head Learning and Teaching in the School of Engineering, University of Tasmania and has responsibility for coordinating all of the university's engineering undergraduate and honours programs. He conducts research in the areas of structural health monitoring, structural dynamics, buckling and acoustics, and he teaches structural mechanics, vibrations and finite element analysis. He has published around 70 peer-reviewed journal articles, supervised 12 PhDs to completion, and has been a core member of the University of Tasmania's high-speed catamaran research group collaboration with shipbuilder Incat for around 25 years.

Prof Spiridon Penev is Professor at the Department of Statistics, UNSW Sydney, Australia. He is teaching Statistics courses at both undergraduate and postgraduate level and is supervising honours and postgraduate students on several topics including machine learning and deep learning.
His research is in statistical methodology in general and in nonparametric statistics with emphasis on wavelet methods, in particular.

Dr Daniele Dessi is Senior Research Scientist at the Institute of Marine Engineering at the National Research Council of Italy where he is responsible for the Laboratory of Dynamics and Diagnostics of Structures. He holds a Master of aeronautics and a PhD in Aerospace Engineering from the University of Rome "La Sapienza". His previous experience includes fluid-structure interaction, hydroelasticity, structural health monitoring, scaled model testing and energy harvesting from vibrations and since 2006 he has been a member of the International Ship and Offshore Structures Congress (ISSC).

Prof Giles Thomas is currently the BMT Chair of Maritime Engineering at University College London. A naval architect, he has over 28 years of research and development experience in the fields of hydrodynamics and fluid-structure interaction working with a large range of industry partners. He is currently the Technical Lead for the EU project SEDNA: Safe maritime operations under extreme conditions: the Arctic case.

UTILIZATION OF ARTIFICIAL NEURAL NETWORKS AS A BASIS FOR DECISION SUPPORT SYSTEM

S. Gökçay and **A. Z. Saydam,** Hidroteknik Nautical Design & Piri Reis University, Turkey
M. İnsel, Hidroteknik Nautical Design, Turkey
G. N. Küçüksu, Hidroteknik Nautical Design & Istanbul Technical University, Turkey

SUMMARY

This paper focuses on the development of an artificial neural networks based on board decision support system that can be utilized for energy efficient ship operations. A method of real time fuel consumption rate prediction model is proposed which can be used during the optimisation process of operational parameters of a ship and support the personnel during the decision of different voyage scenarios that can be used in weather routing, slow steaming, trim and ballast optimisation purposes. To determine the highly correlated input parameters with ships' energy consumption, and to tune in the best neural network architecture, sea trials were conducted and acquired data of a double ended ferry were utilised for the training and the validation of the artificial neural network. Developed network was tested and promising results show that the proposed method has high potential of predicting real time fuel consumption rate values accurately which is essential for a reliable decision support system.

1. INTRODUCTION

While more than 80% percent of world trade by volume is transported by ships, and shipping is the most energy efficient mode of transport, global approach to further improve its energy efficiency and effective emission control is needed as sea transport will continue growing parallel to world trade. Increasing energy efficiency of ships has received much interest to reduce operational costs due to highly volatile fuel prices and greenhouse gas emissions due to environmental concerns. Starting from design of the ship or retrofit solutions applied to existing ships can be categorised as technical and adjusting parameters effecting ship power requirement on the voyage such as trim, draught, route, engine parameters, energy expenditure can be categorized as operational measures. These measures are nowadays also forced by authorities like International Maritime Organisation (IMO).

Optimisation of above parameters for an efficient operation relies heavily on environmental factors like weather conditions and operational conditions like loading conditions of the ship which should be considered simultaneously and in a real time manner by the ship operator. Decision support systems may be a promising solution to this challenge which may provide a reliable optimisation tool to find out the best alternative for an efficient ship operation for a voyage.
Decision support systems were developed to support ship bridge operators on their decisions of optimum route and operational parameters selection. These systems are either based on simplified models of ship performance, reinforced models combined with advanced engineering simulations (like CFD) or direct collection of operational data on route and system identification methodology such as neural network to generate the decision suggestions [1].

In recent years, artificial neural network based models are attracting attention owing to the promising potential of including many parameters effecting ship energy efficiency into models. Common practice followed is to establish a highly correlated link between selected ship operational inputs and energy consumption parameter of the ship as an output. Fuel consumption rate is usually selected, which is a good indicator of ship energy consumption, as an output. This link between inputs and output, is obtained by developing a model within black box approach by utilising artificial neural networks. Good practices of applications can be given in the studies: [2], [3] and [4].

Majority of these studies use limited number of input parameters acquired within long data sampling intervals starting from 5 minutes to a day. To the best of authors' knowledge, there is no study that utilises various number of parameters acquired in seconds to capture the sudden changes for more sensitive artificial neural network predictions. Obviously, this is an important issue particularly for the vessels which have frequently changing operational parameters in their operation profile as in the case taken in this study.

The purpose of the paper is to present (i) the case study selected to gather necessary data for ANN model establishment, (ii) the details of the ANN model that can predict real time fuel consumption rate parameter during the voyage (iii) the ability of the model that can be a reliable base to decision support system for more energy efficiently operating ships.

The paper is organised as follows: Methods and input parameter selection is presented in Section 2. In section 3, data set acquisition details are given. Results of the proposed model is discussed in Section 4. General conclusions are derived and recommendations for further studies are given in Section 5.

2. METHODOLOGY

2.1 ARTIFICIAL NEURAL NETWORKS

The ANN applications had grown gradually from the modelling of simple processing elements or neurons to massively parallel neural networks, which led to more sophisticated artificial intelligence and machine learning applications. ANN has proven its applicability in many fields to forecast its systems outputs, especially systems with many nonlinear relationships involved between its elements. These models have facilitated to solve many problems that could not be solved with traditional methods. The ANN ability to model the nonlinear processes allowed its implementation to wide range of diverse applications [3].

ANN finds out the complex relation between input and output information by using connected neurons which are inspired by human brain. The information is processed through network of neurons by utilising weights and biases. Randomly selected weights are assigned at the initialisation phase. During the learning process, the weights for each neuron are changed to minimise the calculated fitting errors. Following this process, relationship between input and output parameters can be determined.

The architecture of neural network plays an important role for the accuracy of the results. According to nature of the problem, appropriate architecture must be selected. As the complexity of the problem increases, selection of multilayer architecture is essential.

Selected architecture for this study has three layers which are named as input, hidden and output layers. The input layer consists of selected 18 input parameters sensitive to energy consumption while the hidden layer consists of 20 neurons. Output layer parameter consists of one output parameter which is fuel consumption rate. The architecture of the selected network is presented in Figure 1. Lines between the nodes show the informational flow from one neuron to the other. Feed Forward Network type is selected where the information flows from input to output without any complex feedback mechanism.

Neural Network is modelled in by developing a code that uses the Neural Network fitting tool included in the deep learning toolbox of a commercial software. The problem is reduced to solving a data fitting problem and input and output parameters were prepared as matrices in accordance with Neural Network fitting tool format. Levenberg-Marquardt backpropagation algorithm is assigned to train the neural network. The algorithm updates weight and bias values according to Levenberg-Marquardt optimization. This method is often the fastest backpropagation algorithm in the toolbox and is highly recommended as a first-choice supervised algorithm, although it does require more memory than other algorithms [5].

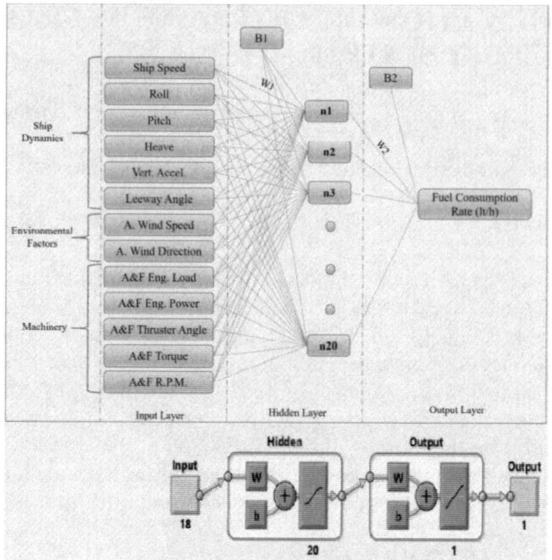

Figure 1: The architecture of used ANN structure.

2.2 SELECTION OF INPUT PARAMETERS

Input parameters are classified into 3 categories. The first group includes parameters related with ship motions that are grouped as "Ship Dynamics". Within this group, 6 parameters are selected, ship speed is the main driver of the ship powering hence energy consumption and fuel consumption. The power required to propel the ship is highly correlated with 3rd power of the ship speed.

Roll, heave, pitch, and vertical acceleration parameters are selected for taking into account the effects of the ship response to the seaway. Translational and rotational motions in seaway directly affect the added resistance of the ship in waves which should be considered for an accurate energy consumption calculation.

Leeway angle is an important parameter while the ship is sailing at cross flow environment like in dominated tides occurring in areas such as straits. As the leeway increases the resistance characteristics and propulsion efficiency of the ship deviates from the calm water conditions.

Apparent wind speed and apparent wind direction are selected to consider the effects of "Environmental Factors". As the wind speed increases, aerodynamic resistance of the hull increases, and the wind direction plays an important role for determining the resistance component of the aerodynamic forces acting on the hull. Also, the ship motions are highly correlated with these two parameters.

As the ship has fore and aft thrusters working simultaneously, thruster angle is selected to capture the effect of the thrust vector hence propulsion efficiency changes in time. Engine load and engine power are selected to determine the effects of machinery parameter combination on fuel consumption while the shaft torque and rpm values are acquired to drive the delivered power variation in time which is highly correlated parameter for the energy consumption.

Since the ship is a double ended ferry, there are aft and fore machinery rooms and the systems are working simultaneously. Therefore, parameters taken in the Machinery part are doubled and named as fore and aft parameters.

The input layer of the proposed artificial neural network architecture consists of these 18 input parameters explained above and can be seen in Figure 1.

3. ANN TRAINING DATASET ACQUSITION

Extensive sea trials were conducted to acquire data that will be used for training and validation of the artificial neural network model. Double ended passenger ferry was selected as a platform for this study. Particulars of the ship is given in Table 1.

Table 1: Ship Particulars

Name:	*ÜSKÜDAR*
IMO No:	9806574
Vessel Type:	Double Ended Passenger Ferry
Owner:	DENTUR AVRASYA GR.
Date Delivered:	2017
Length Overall:	37 m
Beam:	8 m
Draft:	1.6 m
Displacement:	160 mt
Passenger Capacity:	315 Passengers

Operation line of the ship is Uskudar-Besiktas ports which are located opposite sides of the Bosphorus strait. The ship is fitted with fore and aft thrusters working simultaneously. This enables rapid approach and departure manoeuvring of a ship that lead to a significant decrease in elapsed voyage time hence increase in daily number of voyage statistics. Details of the ship is presented in Figure 2.

Within the scope of the testing campaign, sea trials were planned in two phases. The first stage includes measurements during the normal operation at Bosphorus strait, while the second stage was performed during the voyage to the maintenance shipyard, located at Yalova. Trial data subject to this study acquired from the second stage sea trials. The route from Yalova to Istanbul (Harem) is given in Figure 3.

Figure 2: Passenger Ferry Before Sea trial

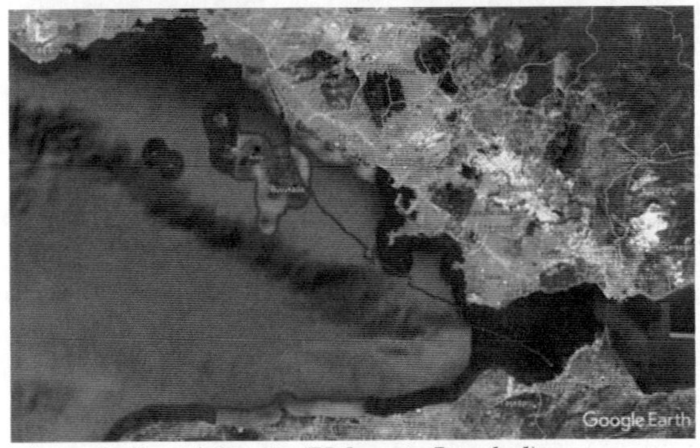

Figure 3: Sea trial route (Yalova to Istanbul)

Before the measurements, measurement equipment was fitted into to the ship during the maintenance works in shipyard. The equipment fitted into the ship, equipment locations and measured parameters (19 parameters) are listed as follows:

- 3D Gyro-DGPS- Accelerometer (Bridge Location)
  - ➢ Ship speed (kn) - GPS
  - ➢ Roll (Deg.)
  - ➢ Pitch (Deg.)
  - ➢ Heave (m)
  - ➢ Vertical Acceleration (m/s^2)
  - ➢ Leeway (Deg.)
- Cameras (Fore & Aft Engine Room)
  - ➢ Fore & Aft Thruster Angle (Deg.)
  - ➢ Fore & Aft Engine Load (%)
  - ➢ Fore & Aft Engine Power (kW)
  - ➢ Fore & Aft Engine Fuel Consumption Rate (lt/h)
- Strain Gauges & RPM Laser Counter (Fore & Aft Engine Room)
  - ➢ Fore & Aft Engine Revolution Speed (rpm)
  - ➢ Fore & Aft Shaft Torque (KNm)
- Wind Anemometer (Mast-Data logged at Bridge)
  - ➢ Apparent Wind Speed (kn)
  - ➢ Apparent Wind Direction (Deg.)

Details of the data acquisition system are schematically presented in Figure 4.

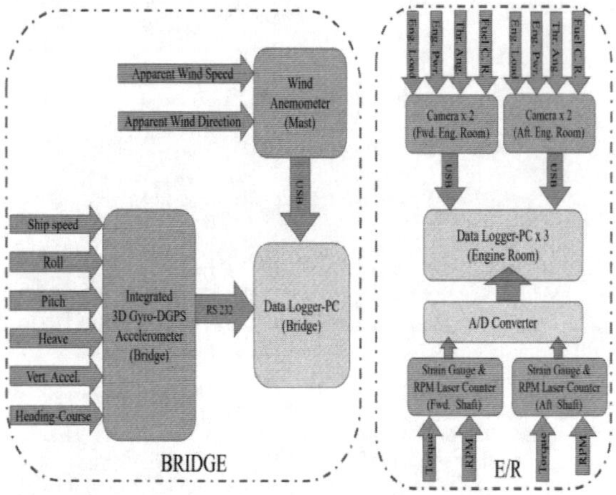

Figure 4: Data Acquisition System Presentation.

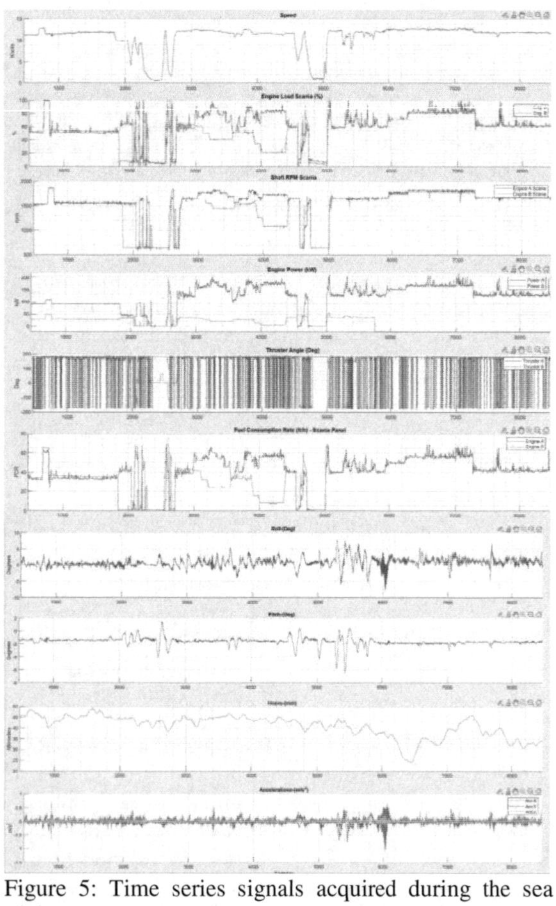

Figure 5: Time series signals acquired during the sea trial.

Ship manoeuvring tests (like turning, stopping and acceleration), maximum speed tests and thruster load sharing tests were performed during the trial in order to gather enough data to train, validate and test the artificial neural network models. Above given parameters were acquired in 1 Hz sampling frequency for 2 hours 10 mins. of voyage time (19x7884 samples were acquired). In Figure 5, the whole acquired data was visualised.

Acquired discrete time domain signals were post-processed and formatted in accordance with the required ANN input. Fuel consumption rate signal was used as output vector in order to validate and evaluate the performance of the model.

4. RESULTS & DISCUSSION

A neural network model is designed by developing a code that uses the Neural Network fitting tool included in the deep learning toolbox provided by a commercial software. The ideal network structure is determined by tuning the number of hidden layers, neurons in the hidden layer and training algorithm through trial and error method. In order to train the network, sea trial data is employed.

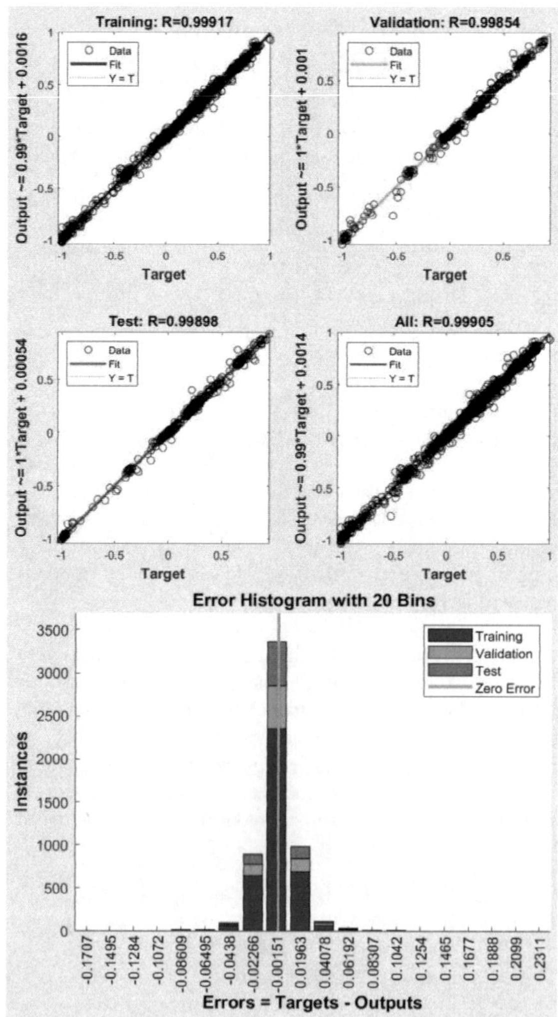

Figure 6: Performance of the Neural Network.

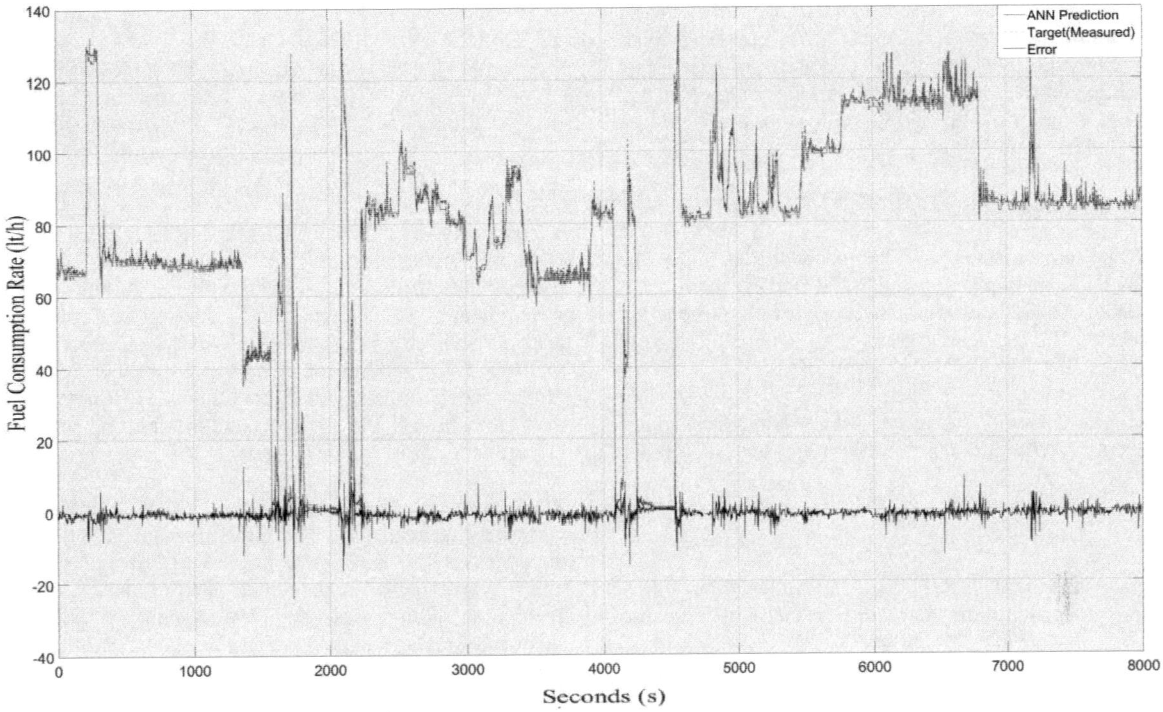

Figure 7: ANN prediction vs measured data, with fitting errors.

Acquired measurement data is randomised to incorporate in the input layer and 5519 samples (70%) are used to train the network while 1183 samples (15%) and another 1183 samples (15%) are used for validation and testing purposes. Levenberg-Marquardt learning algorithm is assigned with learning rate of 0.8. Hyperbolic tangent sigmoid and linear transfer functions are used, and the input parameters are normalised within the range of -1 and 1.

Performance of the proposed artificial neural network structure is presented in Figure 6. The figure contains plots of the training, validation, test and overall fitting regressions and the error histogram of the values calculated for fuel consumption rate output parameter. Targets are the measured values during the sea trial while "output" ordinates correspond to calculated values of employed artificial neural network. R^2 values for the validation are found to be 0.998, 0.997 and 0.997 for training, validation, and test respectively.

As mentioned in former sections, the training of the network is performed by using randomised data in order to prevent over fitting issues. Following to successful fit of the neural network, indicated by higher correlation coefficient values of the regression, saved neural network is employed for finding the output of sequential sea trial data.

Figure 7 presents the predicted fuel consumption rate for the whole voyage by the artificial neural network.

Measured data is also plotted to compare the performance of the data fit. Errors between the prediction and realised values are also plotted for better understanding. Root Mean Square value of the error matrix are calculated as 1.78lt/h that clearly indicates promising result, when the average fuel consumption rate values are considered. Some spikes in error values (just for one second) encountered around 20lt/h values. Obviously, it is due to sudden response of the operator to the throttle and found to be negligible effect on fuel consumption when considering the whole voyage average values.

5. CONCLUSION

Decision support systems are becoming a widely addressed tools with its high potential of supporting ship operators through their response to lowering fuel consumption.

A research was initiated to propose a network architecture with sensitive input parameters to energy consumption. A case study is selected to gather necessary data for ANN model establishment. The details of the ANN model that can predict real time fuel consumption rate parameter during the voyage is presented.

Results of the study show that, the ability of the model that can be a reliable base to decision support system for more energy efficiently operating ships is satisfactory when the measured outputs are compared with the predicted ones.

Application of the proposed model to different ship types and integration of artificial neural network-based fuel consumption rate prediction to a decision support system through state-of-the-art optimisation algorithms may be included in future research.

6. ACKNOWLEDGEMENTS

This research campaign is being conducted under the financial support of The Scientific and Technological Research Council of Turkey (TUBITAK), project no: 3190525. The financial support of the Council is greatly appreciated.

The authors would also like to acknowledge the support of Dentur Avrasya Group for allocating their vessels for the measurements.

7. REFERENCES

1. INSEL, M., GOKCAY, S., SAYDAM, A. Z., 'A decision support system for energy efficient ship propulsion', *Trends and Challenges in Maritime Energy Management, WMUSTUD, Volume 6, Pages 143-155,* 2018.

2. BEŞIKÇI, E. B., ARSLAN O., TURAN O., ÖLÇER, A. I., 'An artificial neural network based decision support system for energy efficient ship operations', *Computers & Operations Research, Volume 66*, 2016.

3. FARAG, Y. B. A., ÖLÇER, A.I., 'The development of a ship performance model in varying operating conditions based on ANN and regression techniques', *Ocean Engineering, Volume 198*, 2020.

4. PETERSEN, J. P., WINTHER, O., & JACOBSEN, D. J., 'A machine learning approach to predict main energy consumption under realistic operational conditions', *Ship Technology Research, 59(1), 64–72*, 2012.

5. MARK HUDSON BEALE, MARTIN T. HAGAN, HOWARD B. DEMUTH, 'Deep Learning Toolbox, Reference Guide', *Mathworks*, 2020.

8. AUTHORS BIOGRAPHY

Serhan Gokcay, Ph.D. holds the current position of technical manager at Hidroteknik nautical design technologies Ltd. He is responsible for hull form development, seakeeping analysis, design performance analysis and full-scale testing. His previous experience includes design and testing of sailing & motor yachts and development of decision support systems. He is also full-time lecturer at Naval Architect and Marine Engineering Department of Piri Reis University.

Ziya Saydam, Ph.D. holds the current position of general manager at Hidroteknik nautical design technologies Ltd. He is responsible for CFD studies. His previous experience includes hull form design of motor vessels & sailing boats and development of decision support systems. He is also full-time lecturer at Naval Architect and Marine Engineering Department of Piri Reis University.

Mustafa Insel, Ph.D. holds the current position of business development manager at Hidroteknik nautical design technologies Ltd. He is responsible for software development and physical testing for new applications. His previous experience includes expert systems, development of decision support systems and full-scale measurements of ships. He is also part time professor at World Maritime University, Malmö.

Gözde Nur Küçüksu holds the current position of research and development engineer at Hidroteknik nautical design technologies Ltd. She is responsible for hull form geometry preparation, potential flow calculations, mesh generation and optimisation studies. She is also a Ph.D. candidate at Istanbul Technical University Naval Architecture program.

